Julian Schwinger

The Physicist, the Teacher, and the Man

Julian Schwinger

The Physicist, the Teacher, and the Man

editor

Prof. Y. Jack Ng
University of North Carolina at Chapel Hill

Published by

World Scientific Publishing Co. Pte. Ltd.
5 Toh Tuck Link, Singapore 596224
USA office: 27 Warren Street, Suite 401-402, Hackensack, NJ 07601
UK office: 57 Shelton Street, Covent Garden, London WC2H 9HE

Library of Congress Cataloging-in-Publication Data
Julian Schwinger: the physicist, the teacher, and the man / editor, Y. Jack Ng.
    p. cm.
    Includes bibliographical references.
    ISBN-13 978-981-02-2531-5 -- ISBN-10 981-02-2531-8
    ISBN-13 978-981-02-2532-2 (pbk) -- ISBN-10 981-02-2532-6 (pbk)
    1. Schwinger, Julian Seymour, 1918–1994 -- Congresses. 2. Physics -- History -- Congresses.
3. Physicists -- United States -- Biography -- Congresses. I. Ng, Y. Jack (Yee Jack)
QC16.S29J85 1996
530'.092--dc20   95-45970
                                                                                         CIP

British Library Cataloguing-in-Publication Data
A catalogue record for this book is available from the British Library.

Copyright © 1996 by World Scientific Publishing Co. Pte. Ltd.

All rights reserved. This book, or parts thereof, may not be reproduced in any form or by any means, electronic or mechanical, including photocopying, recording or any information storage and retrieval system now known or to be invented, without written permission from the publisher.

For photocopying of material in this volume, please pay a copying fee through the Copyright Clearance Center, Inc., 222 Rosewood Drive, Danvers, MA 01923, USA. In this case permission to photocopy is not required from the publisher.

The marble index of a mind forever
Voyaging through strange seas of thought, alone.

          Wordsworth

PREFACE

In the post-quantum-mechanics era, few physicists, if any, have matched Julian Schwinger in contributions to and influence on the development of physics. A towering giant in theoretical physics, Schwinger left his indelible mark on such diverse fields as quantum mechanics, quantum field theory, electrodynamics, nuclear physics, statistical mechanics, atomic physics, elementary particle physics, gravity, and mathematical physics. On July 16, 1994, he succumbed to pancreatic cancer at the age of seventy-six. This book is a collection of talks by some of his contemporaries and his former students in memory of him. In these pages one can get a glimpse of Julian Schwinger, the physicist, the teacher, and the man.

Of him, P. C. Martin and S. L. Glashow, two of his former students, wrote, "Rare is the theoretical physicist who makes repeated and varied contributions apart from the throng; rarer still one who not only contributes but sets standards and priorities single-handedly. Julian Seymour Schwinger ... was such an individual.... His ideas, discoveries, and techniques pervade all areas of theoretical physics."

Schwinger was born in New York City on February 12, 1918. Fiercely independent, as a child he taught himself physics and mathematics by reading books and journals in nearby libraries. With I. I. Rabi as his mentor, he received his Ph.D. in physics from Columbia University at the age of twenty-one. After postdoctoral studies at Columbia and Berkeley, he worked on radar for the Allied war effort at the Radiation Laboratory at M.I.T. At the end of World War II, he accepted a professorship from Harvard University where he conducted the work which would win him a Nobel Prize. In spring 1971, he moved to UCLA, where he became a University Professor.

Schwinger shared the first Albert Einstein Prize with the mathematician Kurt Gödel in 1951. He was awarded the newly created National Medal of Science in 1964. In 1965, he shared the Nobel Prize in Physics with R. Feynman and S. Tomonaga for their "fundamental work" on quantum electrodynamics with "deep-ploughing consequences for the physics of elementary particle." The theory of QED has withstood the test of time; it is arguably the most accurate theory ever devised by mankind. The trio's independent work on QED (and that of F. Dyson among others) made use of a new concept in quantum field theory known as renormalization. In his appraisal of Schwinger's contributions to physics, C. N. Yang observed, "Renormalization was one of the great peaks of the development of fundamental physics in this century. Scaling this peak was a difficult enterprise.... Many many people can climb the peak now. But, the person who first conquered the peak was Julian Schwinger."

In addition to QED, Schwinger made fundamental contributions to many areas of physics. He laid down the foundation for much of quantum field theory as we understand it today. Schwinger introduced operator and functional techniques, Euclidean and finite temperature field theories, proper-time methods, and strong field techniques. His insightful papers on anomalies (he discovered the axial-vector anomaly), commutations of currents, commutations of stress tensor components, spin-statistics connections, and the TCP theorem are classic works in quantum field theory. Schwinger's prescient contributions to elementary particle physics proved to be crucial. It was he who first realized the need for more than one type of neutrino. The seminal idea and framework of electroweak unification was prefigured in his 1957 *Annals of Physics* paper. About this spectacular paper he wrote, "A speculative paper that was remarkably on target: VA weak interaction theory, two neutrinos, charged intermediate vector meson, dynamical unification of weak and electromagnetic interactions, scale invariance, chiral transformations, mass generation through vacuum expectation value of scalar field."

Among Schwinger's other contributions, his illuminating work on chiral dynamics and effective Lagrangians greatly enhanced our understanding of low-energy hadronic physics. In scattering theory, he showed the world how to use the concept of effective range. The techniques he developed in 1961 for treating quantum systems away from equilibrium are used nowadays in studies of microelectronic devices, quark-gluon plasmas, and cosmology. Some of his works are even of great interest to mathematicians, who speak of the Kubo-Martin-Schwinger analyticity properties of the finite-temperature Green's function, and of the Schwinger conjecture in connection with his work on multi-electron atoms. His name is associated with many ideas and techniques in physics: the Schwinger-Dyson equation, the Schwinger-Feynman parametrization (erroneously called the Feynman parametrization in the literature), the Schwinger mechanism (for generating mass for fermions), the Schwinger model, the Schwinger term, the Lippmann-Schwinger equation, the Rarita-Schwinger field, the Schwinger proper-time method, the Schwinger-DeWitt technique (for calculating Green's functions in curved spaces), the Schwinger closed time path formalism, the Schwinger-Jordan representation (of angular momentum in terms of two oscillators), and of course, the Schwinger action principle. The list goes on and on.

But, despite this impressive list of path-breaking achievements, Schwinger suffered two major near misses. By 1957, he had almost all the ingredients to construct the SU(2) × U(1) electroweak theory. Yet he failed to follow up on his own idea of electroweak unification. Wrong experiments at that time obviously played a major role in this failure. But Schwinger had another reason:

"Concerning the idea of unifying weak and electromagnetic interactions, Rabi once reported to me: 'They hate it'." This time, Schwinger's trust in his former mentor's intuition was misplaced. Fortunately for physics, he suggested the problem to Glashow for further investigation. Schwinger must also have been quite disappointed with himself for not coming up with the idea of supersymmetry and supergravity. After all, it was he who had introduced the multispinor description of particle fields. And, as he wrote in his 1979 *Annals of Physics* paper, "[That] description ... provides a unification of all spins and statistics. As such, it makes quite transparent the existence of transformations that alter the particle spin by 1/2, with the accompanying Fermi-Bose transformation of statistics." This miss on the part of Schwinger was particularly ironic in view of the fact that supergravity makes crucial use of a field that bears his name: the Rarita-Schwinger field.

Schwinger was a perfectionist, not only as a researcher, but also as a lecturer. Throughout his long career he prepared every lecture meticulously. Hearing him lecture was commonly likened to listening to a virtuoso soloist-composer performing an original and brilliant composition. His lecture notes were widely disseminated throughout the physics community. For his distinguished teaching, he was honored with the Sigma Xi Award in 1986. Perhaps even more impressive was Schwinger's success as a mentor. In all, he directed more than seventy doctoral theses. Among his former students are two Nobel laureates: B. Mottelson and S. Glashow. (An interesting note: the doctor who diagnosed Schwinger of his terminal cancer was one of his academic great-grandsons.)

As a person, Schwinger was gentle, a bit shy, intensely private, and extremely cultivated. According to Yang, he "epitomized the cultured perfectionist and the quiet inward-looking gentleman." D. Saxon, Schwinger's coauthor of their influential book on waveguides, observed that Schwinger "read widely in history and novels...."

Small talk was not his forte; his sense of humor was too subtle for that, but he was a great listener and a deep and provocative thinker." An unassuming man, Schwinger was "modest about everything except his physics," said Clarice Schwinger, his devoted wife of forty-seven years. (Perhaps she could have included his cars among the exceptions — it is said that he used to commute to work in a Cadillac.) Schwinger was a music lover and played the piano. Of his piano playing, he once said, "If something is worth doing, it is worth doing badly."

After his untimely death, his friends and former students held three symposia to commemorate his remarkable achievements. A. Klein organized the first one at Drexel University on September 11, 1994. The second one was held at UCLA on October 22, 1994, organized by Saxon. The American Physical Society held a Schwinger memorial session in its joint meeting with the American Association of Physics Teachers in Washington, D.C., on April 20, 1995; it was cochaired by Y. J. Ng and R. Peccei. This book contains all the talks given in the three gatherings. At Drexel, after Klein's elegant comment about Schwinger, Dyson gave a beautiful reading of Schwinger's last public lecture, in which Schwinger summed up his legendary applications of Green's functions. Reflecting the diversity of his students' specialties, the other two speakers, B. DeWitt and W. Kohn, both former students of Schwinger, gave an account of their respective fields of expertise, curved spacetime propagators and density functional theory. At UCLA, all the speakers, Saxon, Martin, K. Johnson, S. Deser, and R. Finkelstein, reminisced on their long interaction and association with Schwinger, and on his various contributions to physics. At the Joint APS-AAPT meeting, after Ng's introductory remarks, H. Feshbach spoke of Schwinger's early years and his contributions to nuclear physics. L. Brown highlighted some of Schwinger's mathematical techniques, which are now a part of every theorist's arsenal. Glashow then gave a gracious discussion of Schwinger's decisive contribution to the development of the electroweak theory. K. Milton covered Schwinger's works at UCLA and discussed source theory, which was the master's preoccupation for his last twenty-seven years. In the concluding talk, Yang gave a short, frank, and eloquent appraisal of Schwinger's work and personality.

Schwinger would have been happy and proud to hear what his friends and former students said about him. To his memory we now dedicate this book, in witness of our appreciation and affection for him.

Y. Jack Ng

Chapel Hill, NC
August, 1995

CONTENTS

Preface vii

1. Recollections of Julian Schwinger 1
   A. Klein

2. Schwinger's Response to the Award of an Honorary
   Degree at Nottingham; 9
   Schwinger's "The Greening of Quantum Field Theory:
   George and I" 13
   F. Dyson

3. The Uses and Implications of Curved-Spacetime
   Propagators: A Personal View 29
   B. DeWitt

4. Overview of Density Functional Theory 61
   W. Kohn

5a. Julian Schwinger Memorial Tribute 79
    D. Saxon

6. Julian Schwinger — Personal Recollections 83
   P. C. Martin

7. Julian Schwinger — Personal Recollections 91
   K. Johnson

8. Julian Schwinger — Personal Recollections 99
   S. Deser

9. Julian Schwinger: The QED Period at Michigan and
   the Source Theory Period at UCLA   105
   R. Finkelstein

5b. Julian Schwinger Memorial Tribute   111
    D. Saxon

10. Schwinging a Sorcerer's Wand: Julian and I   117
    Y. J. Ng

11. Julian Schwinger — Reminiscences and
    Nuclear Physics   121
    H. Feshbach

12. An Important Schwinger Legacy: Theoretical Tools . .   131
    L. Brown

13. The Road to Electroweak Unification   155
    S. Glashow

14. Julian Schwinger: Source Theory and the UCLA
    Years — from Magnetic Charge to the Casimir Effect . 161
    K. A. Milton

15. Julian Schwinger   175
    C. N. Yang

APPENDIX 1:   SCHWINGER'S DOCTORAL
              STUDENTS   181

APPENDIX 2:   PUBLICATIONS OF
              JULIAN SCHWINGER   183

ABOUT THE EDITOR   195

Julian Schwinger as a teenager, circa 1931.

RECOLLECTIONS OF JULIAN SCHWINGER

Abraham Klein

University of Pennsylvania and Drexel University

Abstract

The recollections that follow were delivered as the introductory remarks at a special session in memory of Julian Schwinger that took place on Sunday, Sept. 11, 1994 at Drexel University following a symposium on the quantum-classical correspondence. The author of these remarks had the privilege of chairing a session in which the speakers were Walter Kohn, Bryce DeWitt, and Freeman Dyson.

We are here to honor the memory of Julian Schwinger, one of the great masters of theoretical physics of this century, who died on Saturday, July 16, at the age of seventy-six. I am lucky to have been one of his many students. I have given an account of my relationship with him as part of a short, published memoir, also distributed by e-mail. For a generally balanced, scientific biography, one can hardly do better than the chapter on Schwinger in Sam Schweber's recent book, *QED and the Men Who Made It, Dyson, Feynman, Schwinger and Tomonaga*.

We are fortunate to have Freeman Dyson as one of our speakers. We are equally fortunate to have as our other speakers two of Julian Schwinger's first generation of graduate students, who have gone on to distinguished careers in quite diverse disciplines. Before passing the word to them, I beg their indulgence for interposing a largely personal account of my interactions with and opinions concerning Julian Schwinger. However, I precede that account with a quote from the appreciation of Julian that formed the preface to the book of essays in honor of his sixtieth birthday:

> His work during the forty-four years preceding his sixtieth birthday extends to almost every frontier of modern theoretical physics. He has made far-reaching contributions to nuclear, particle, and atomic physics, to statistical mechanics, to classical electrodynamics, and to general relativity. Many of the mathematical techniques he developed can be found in every theorist's arsenal.... He is one of the prophets and pioneers in the uses of gauge theories.... Schwinger's influence, however, extends beyond his papers and books. His course lectures and their derivatives constitute the substance of graduate physics courses throughout the world, and in addition to directing about seventy doctoral theses, he is now the ancestor of at least four generation of physicists.... The influence of Julian Schwinger on the physics of his time has been profound.

I need hardly add that I concur with every word of this quote.

I arrived in Cambridge in January, 1947 and left in June, 1955, to take up a position at the University of Pennsylvania. In 1947, I didn't know even the most elementary quantum mechanics. When I left, I was considered something of an expert in the Dyson-Schwinger formulation of field theory and related matters. In fact this so-called expertise played an essential role in my research for almost another decade, until I finally found my personal, if minor, voice.

How much of that education did I owe to Julian Schwinger? My answer is that I owed a great deal, but if at the same time you ask me how many *personal* interactions I had with Julian, during the 8.5 years I passed in Cambridge, that were intellectually decisive for my career, I can honestly recall only three. The first occurred in early 1950, when he explained to me the essential ideas of the only part of my thesis that turned into a paper, my first. The second was a year later, when as an instructor in the physics department and one of two Schwinger assistants, he gave me the notes of a beautiful new method he had developed for doing the Lamb shift calculation, that became the starting point for a three-sided collaboration involving Schwinger, Robert Karplus — an assistant professor and the other assistant and myself. This collaboration resulted in two significant publications, one on the hyperfine interval in Hydrogen and the other on the Lamb shift proper. The third time was four years later, during my last year in the Society of Fellows at Harvard. In the meantime I had become an independent physicist, but from time to time went to see Julian, as a courtesy, to tell him about my current research. On this particular occasion he showed me that if I were to incorporate a technique of effective Lagrangians and functional differentiation into my study of low energy theorems in field theory, the work would be vastly improved, as indeed it was. I think the above speaks volumes for the way that Schwinger interacted with the young people around him, at least during those early years. If he thought you needed help, he did his best to provide it. Otherwise it was *laissez faire*. I shall return to this question below.

Concerning Schwinger's brilliance as a lecturer, it is widely acknowledged that for many years he was almost in a class by himself. Though he seemed to move rapidly, generally it was possible to take notes and follow the thread of the argument, because he repeated ideas two or even three times (but never in the same words). On the other hand he did tend to smooth over difficulties and it was clear that he didn't encourage questions, so that none was ever asked, at least during his classroom lectures. I attended every course, every set of special lectures, and every seminar that Julian gave on field theory as long as I remained in Cambridge.

I would like at this point to correct a glaring error made by James Gleick in his generally excellent biography of Feynman. Because the methods used by Schwinger in his early papers on QED were superseded by the Feynman formulation, mainly as a result of Freeman Dyson's work, Gleick claims that by the early fifties, the young physicists around Schwinger found themselves at a distinct disadvantage. In fact, it was precisely during this period that Schwinger developed, lectured on and published his functional version of the Feynman-Dyson formulation. I believe I saw a derivation of the so-called Bethe-Salpeter equation in a lecture by Julian before I ever read the famous paper. (The credit for the discovery of this equation should probably go to Y. Nambu, who published the first paper on the subject.) In addition, Karplus, in consequence of his earlier work at the Institute for Advanced Study, was an expert on the evaluation of Feynman diagrams, and the rest of us were not exactly illiterate. In addition to the work mentioned above, during this period Karplus and I, using Schwinger's methods as a starting point, developed and applied the Bethe-Salpeter formalism to positronium. There followed theses on this and other problems in QED by Paul Martin, Tom Fulton, Roger Newton, and Charles Zemach. One can even argue that this was the preeminent work on the applications of QED that was done during these years.

Of course there was much, much more going on with Julian that had nothing to do with the young people around him. In 1958, he edited a collection of the classics of field theory published up to about 1955. Among a selection of 34 papers, he included six of his own of which four were written after 1950. These were his beautiful papers on gauge invariance and vacuum polarization, the first two papers of the series he wrote on the quantum action principle, and the short paper in the *Proceedings of the National Academy of Sciences* that laid the foundations of the Schwinger-Dyson formulation. (Of course the collection contains all the old classics in the field as well as the great papers of Feynman and Dyson. Perhaps you will forgive me for adding that he was kind enough to include the paper on positronium by Karplus and Klein.)

Schweber properly identifies the decade from 1945-1955 as the glory years of Schwinger's supreme achievement and influence. Prior to that he was simply the most promising young theorist in the country, whose singular abilities were recognized by all who interacted with him. In the years following the late fifties, he continued for a long time to write deep papers and to produce outstanding students. For example, of his two students who won Nobel prizes, Ben Mottelson was my contemporary, but Shelly Glashow, who was pointed in the right direction by Julian, belongs to the next decade. However, Julian's mantle as the supreme intellectual leader of the field of elementary particle physics passed on to others, as was inevitable. He continued to produce interesting and provocative physics to the very end.

Eugene Wigner is credited with the observation that every physicist has at least two faults. I would like to address what is purported to be one of Julian Schwinger's faults, namely his way of dealing with graduate students. The charge is that he assigned thesis topics in which he had, at best, marginal interest, and that in any event, you didn't get much help after that. The first charge is partly true, but there were also some famous theses, such as the ones which produced the Lippmann-Schwinger equation and the Walter Kohn variational principle. During my time as a junior faculty member, there were very good theses by Paul Martin and Roger Newton. On the other hand, my own thesis was less than mediocre, and the thesis of Stan Deser, recent co-winner of the Dannie Heineman award for his work in general relativity, was also unimpressive. This adds an intriguing element to our discussion, namely the relatively poor correlation between the quality of theses and how careers developed later.

My aim here is to defend or at least rationalize Julian's behavior in these matters, since in my opinion, all the published discussions miss something. There was first of all the pressure of numbers. During the time that I was aware of things, Julian never said no to any student who had been certified by the department to do theoretical physics. During my thesis time, he had ten graduate students. The only other active tenured theorists in the department were Wendell Furry and John van Vleck, both outstanding physicists and wonderful human beings. As I recall Wendell had two students and Van a few more. Most of us, however, were dying to work for the new star in the firmament. But you had to have the courage to ask him to accept you and the courage to ask for a topic, since already in my generation most of us knew too little to be able to pick our own thesis topic. I know that my own courage almost failed on both counts.

Why did we see so little of him once we began work? The stories of how hard it was to get to see him once you decided it was absolutely necessary translate, in practice, that *sometimes* you had to wait up to a week before your turn came. But that does not explain why the average interval between audiences was three months for me, longer for some, shorter for others. My answer, which I believe represents part of the general opinion of his students, is that we were so in awe and had so much respect for the value of his time, as opposed to the value of our own, that we felt it necessary to exhaust all other resources available to us, the literature, our fellow graduate students, and our own efforts, before we went to see Julian. In thinking about those times, I have come to realize that after the short meeting during which I received my first research topic, though I remained in awe of his abilities, I was never again afraid of him, because no matter how poor the quality of the work I had done, he never tried to destroy my ego. I could see that he took my concerns seriously and did his best to come up with useful advice. Only about half the research topics he suggested to me were any good or at least any good for me. Some of his students thought everything should work out perfectly and became and remained angry when it didn't. I think that those of us who were more realistic in our expectations fared better personally. He also took seriously his basic responsibility to order us according to promise and this explains, but only in part, why bad theses sometimes led to good careers.

I want only to add that once I got to know him, I never found Julian's behavior to me to be other than kind and generous. I am and shall remain profoundly indebted to him. I considered him a good friend, but for reasons that have to do both with my personality and his, I was never one of his close friends. I mourn his passing.

PRELIMINARY REMARKS

Freeman Dyson
Institute for Advanced Study

I was lucky to spend three days with Julian Schwinger and his wife Clarice in July 1993, shortly before he got sick with his last illness. We were at the University of Nottingham in England to celebrate the two-hundredth birthday of George Green. Julian was in excellent spirits and enjoyed the festivities as much as I did. He gave two talks, a short talk in response to the award of an honorary degree, and a long talk at the Green celebration. I decided it would make no sense to speak today about my personal memories of Julian, memories which are dim and unreliable after forty years. Instead I will give you the authentic Julian, as we saw and heard him in Nottingham only a year ago. So here are his two talks, first the short one and then the long one. He told me that the short one cost him far more time and effort to prepare. The protocol required him to say what he had to say in less than four minutes. As you will hear, he said more in four minutes than most Commencement speakers say in one hour.

(Prof. Dyson then went on to give a beautiful reading of Schwinger's two talks prepared for the bicentennial celebration of George Green's birthday.)

SCHWINGER'S RESPONSE TO THE AWARD OF AN HONORARY DEGREE AT NOTTINGHAM

On behalf of my colleague, and myself, I give deep thanks, to the University of Nottingham, and to the people who arranged this outstanding event. My congratulations to the recipients of degrees in various branches of Science, and my best wishes to them, along with their accompanying parents and friends.

The Degree Ceremony is a modern version of a medieval rite that seemed to confer a kind of priesthood upon its recipients, thereby excluding all others from its inner circle. But that will not do for today. Science, with its offshoot of Technology, has an overwhelming impact upon all of us. The recent events at Wimbledon invite me to a somewhat outrageous analogy. Very few of us, indeed, are qualified to step onto centre court. Yet thousands of spectators gain great pleasure from watching these talented specialists perform. Something similar should be, but generally is not, true for the relationship between practitioners of Science and the general public. This is much more serious than not knowing the difference between 30 all and deuce. Science, on a big scale, is inevitably intertwined with politics. And politicians have little practice in distinguishing between, say, common law and Newton's law. It is a suitably educated public that must step into the breach. This has been underlined lately by Minister Waldegrave's cry for someone to educate him about the properties of the Higgs boson, to be rewarded by a bottle of champagne. Any member of the educated public could have told him that the cited particle is an artifact of a particular theoretical speculation, and the real challenge is to enter uncharted waters to see what is there. The failure to do this will inevitably put an end to Science. A society that turns in on itself has sown the seeds of its own demise. Early in the 16th century, powerful China had sea-going vessels exploring to the west. Then a signal came from new masters to return and destroy the ships. It was in those years that Portuguese sailors entered the Indian Ocean. The outcome was 400 years of dominance of the East by the West.

There are other threats to Science. A recent bestseller in England *Understanding the present*, has the subtitle *Science and the soul of Modern Man*. I shall only touch on the writer's views toward quantum mechanics, surely the greatest intellectual discovery of the 20th Century. First, he complains that the new physics of quantum mechanics tosses classical physics into the trash bin. This I would dismiss as mere technical ignorance; the manner in which classical and quantum mechanics blend into each other has long been established. Second, the author is upset that its theories can't be understood by anyone not mathematically sophisticated and so must be accepted by most people on faith. He is, in short, saying that there is a priesthood. Against that I pose my own experience in presenting the basic concepts of quantum mechanics to a class of American high school students. They understood it, they loved it. And I used no more than a bit of algebra, a bit of geometry. So: catch them young; educate them properly; and there are no mysteries, no priests. It all comes down to a properly educated public.

THE GREENING OF QUANTUM FIELD THEORY: GEORGE AND I*

Julian Schwinger

University of California, Los Angeles, CA 90024-1547

The young theoretical physicists of a generation or two earlier subscribed to the belief that: If you haven't done something important by age 30, you never will. Obviously, they were unfamiliar with the history of George Green, the miller of Nottingham.

Born, as we all know, exactly two centuries ago, he received, from the age 8, only a few terms of formal education. Thus, he was self-educated in mathematics and physics, when in 1828, at age 35, he published, by subscription, his first and most important work: "An Essay on the Applications of Mathematical Analysis to the Theory of Electricity and Magnetism." The Essay was dedicated to a noble patron of the "Sciences and Literature," the Duke of Newcastle. Green sent his own copy to the Duke. I do not know if it was acknowledged. Indeed, as Albert Einstein is cited as effectively saying, during his 1930 visit to Nottingham, Green, in writing the Essay, was years ahead of his time.

---

*Lecture of July 14, 1993, at Nottingham

There are those who cannot accept that someone, of modest social status and limited formal education, could produce formidable feats of intellect. There is the familiar example of William Shakespeare of Stratford on Avon. It took almost a century and a half to surface, and yet another century to strongly promote, the idea that Will of Stratford could not possibly be the source of the plays and the sonnets which had to have been written by Francis Bacon. Or was it the earl of Rutland? Or perhaps it was William, the sixth earl of Derby? The most recent pretender is Edward deVir, seventeenth earl of Oxford, notwithstanding the fact that he had been dead for 12 years when Will was put to rest.

I have always been surprised that no one has suggested an analogous conspiracy to explain the remarkable mathematical feats of the miller of Nottingham. So I invented one.

Descended from one of the lines of the earl of Nottingham was the branch of the earls of Effindham, which was separated from the Howards in 1731. The fourth holder of the title died in 1816, with apparently no claimant. In that year, George Green, age 23, could well have reached the maturity that led, 12 years later, to the publication of the Essay. And what of the remarkable fact that, in the same year that the earldom was revived, 1837, George Green graduated fourth wrangler at Cambridge University?

The conspiracy at which I hint darkly is one in which I believe quite as much as I think Edward deVir is the real Shakespeare.

I consider myself to be largely self-educated. A major source of information came from my family's possession of the *Encyclopedia Brittanica Eleventh Edition*. I recently became curious to know what I might have, and probably did, learn about George Green, some 65 years before.

There is no article detailing the life of George Green. There are, however, four brief references that indicate the wide range of Green's interests.

First, in the article Electricity, as a footnote to the description of Lord Kelvin's work, is this:

> In this connexion the work of George Green (1793-1841) must not be forgotten. Green's "Essay on the application of mathematical analysis to the theories of electricity and magnetism," published in 1828, contains the first exposition of the theory of potential. An important theorem contained in it is known as Green's theorem, and is of great value.

It was, of course, Lord Kelvin, or rather William Thomson, who rescued Green's work from total obscurity.

Then, in the article Hydromechanics, after several applications of Green's transformation, which is to say, the theorem, there appears, under the heading "The Motion of a Solid through a Liquid":

> The ellipsoid was the shape first worked out, by George Green, in his "Research on the vibration of a pendulum in a fluid medium" (1833).

On to the article Light under the heading "Mechanical Models of the Electromagnetic Medium." After some negative remarks about Fresnel, one reads:

> Thus, George Green, who was the first to apply the theory of elasticity in an unobjectional manner ...

This is the content of "On the Laws of Reflexion and Refraction of Light" (1837).

Finally, the paper "On the Propagation of Light in Crystallized Media" (1839) appears in the Brittanica article Wave as follows:

> The theory of waves diverging from a center in an unlimited crystalline medium has been investigated with a view to optical theory by George Green.

The word "propagation" is a signal to us that, in little more than 10 years, George Green had significantly widened his physical framework. From the static three-dimensional Green function that appears in potential theory, he had arrived at the concept of a dynamical, four-dimensional Green function. It would be invaluable a century later.

To continue the saga of George Green and me — my next step was to trace the influences of George Green on my own works. Here I spent no time over ancient documents. I went directly to a known source: THE WAR.

I presume that in Britain, unlike the United States, *the War* has a unique connotation. Apart from a brief sojourn in Chicago, to see if I wanted to help develop The Bomb — I didn't — I spent the War years helping to develop microwave radar. In the earlier hands of the British, that activity, famous for its role in winning the Battle of Britain, had begun with electromagnetic radio waves of high frequency, to be followed by very high frequency, which led to very high frequency, indeed.

Through those years in Cambridge (Massachusetts, that is), I gave a series of lectures on microwave propagation. A small percentage of them is preserved in a slim volume entitled *Discontinuities in Waveguides*. The word *propagation* will have alerted you to the presence of George Green. Indeed, on pages 10 and 18 of an introduction there are applications of two different forms of Green's identity.

Then, on the first page of Chapter 1, there is Green's function, symbolized by G. In the subsequent 138 pages the references to Green in name or symbol are more than 200 in number.

As the War in Europe was winding down, the experts in high power microwaves began to think of those electric fields as potential electron accelerators. I took a hand in that and devised the microtron which relies on the properties of relativistic energy. I have never seen one, but I have been told that it works. More important and more familiar is the synchrotron.

Here I was mainly interested in the properties of the radiation emitted by an accelerated relativistic electron. I used the four-dimensionally invariant proper time formulation of action. It included the electromagnetic self-action of the charge, which is to say that it employed a four-dimensionally covariant Green's function. I was only interested in the resistive part, describing the flow of energy from the mechanical system into radiation, but I could not help noticing that the mechanical mass had an invariant electromagnetic mass added to it, thereby producing the physical mass of an electron. I had always been told that such a union was not possible. The simple lesson? To arrive at covariant results, use a covariant formulation, and maintain covariance throughout.

Quantum field theory, or more precisely, quantum electrodynamics, was forced from childhood into adolescence by the experimental results announced at Shelter Island early in June, 1947. The relativistic theory of the electron created by Dirac in 1928 was wrong. Not very wrong, but measurably so.

A few days later, I left on a honeymoon tour across the United States. Not until September did I begin to work on the obvious hypothesis that electrodynamic effects were responsible for the experimental deviations, one on the magnetic moment of the electron, the other on the energy spectrum of the hydrogen atom.

Although a covariant method was in order, I felt I could make up time with the then more familiar non-covariant methods of the day. By the end of November I had the results. The predicted shift in magnetic moment agreed with experiment. As for the energy shift in hydrogen, one ran into an expected problem.

Consider the electromagnetic momentum associated with a charge moving at constant speed. The ratio of that momentum to the speed is a mass — an electromagnetic mass. It differs from the electromagnetic mass inferred from the electromagnetic energy. Analogously, the magnetic dipole moment inferred for an electron moving in an electric field is wrong. Replacing it by the correct dipole moment leads to an energy level displacement that was correct in 1947, and remains correct today at that level of accuracy as governed by the fine structure constant.

I described all this at the January 1948 meeting of the American Physical Society, after which Richard Feynman stood up and announced that he had a relativistic method. Well, so did I, but I also had the numbers. Indeed, several months later, at the opening of the Pocono Conference, he ran over to me, shook my hand, and said "Congratulations, Professor! You got it right," which left me somewhat bewildered. It turned out he had completed his own calculation of the additional magnetic moment. Later we compared notes and found much in common.

Unfortunately, one of the things we shared was an incorrect treatment of low energy photons. Nothing fundamental was involved; it was a matter of technique in making a transition between two different gauges. But, as in American politics these days, the less important the subject, the louder the noise. When that lapse was set right, the result of 1947 was regained. Incidentally, even Lord Rayleigh once made a mistake. That's one reason for its being called the Rayleigh-Jeans law.

To keep to the main thrust of the talk — the evolution of Green's function in the quantum mechanical realm — I move on to 1950, and a paper entitled "On Gauge Invariance and Vacuum Polarization."

This paper makes extensive use of Green's functions, in a proper-time context, to deal with a variety of problems: non-linearities of the electromagnetic field, the photon decay of a neutral meson, and a short, but not the shortest derivation of the additional electron magnetic moment. The latter ends with the remark that "The concepts employed here will be discussed at length in later publications." I cannot believe I wrote that.

The first, rather brief, discussion of those concepts appeared in a pair of 1951 papers, entitled "On the Green's Functions of Quantized Fields." One would not be wrong to trace the origin of today's lecture back 42 years to these brief notes. This is how paper I begins:

> The temporal development of quantized fields, in its particle aspect, is described by propagation functions, or Green's functions. The construction of these functions for coupled fields is usually considered from the viewpoint of perturbation theory. Although the latter may be resorted to for detailed calculations, it is desirable to avoid founding the formal theory of the Green's functions on the restricted basis provided by the assumption of expandability in powers of the coupling constants. These notes are a preliminary account of a general theory of Green's functions, in which the defining property is taken to be the representation of the fields of prescribed sources.
>
> We employ a quantum dynamical principle for fields which has been described in the 1951 paper entitled "The Theory of Quantized Fields." This (action) principle is a differential characterization of the function that produces a transformation from eigenvalues of a complete set of commuting operators on one space-like surface to eigenvalues of another set on a different surface.

In one example of a rigorous formulation, Green's function, for an electron-positron, obeys an inhomogeneous Dirac differential equation for an electromagnetic vector potential that is supplemented by a functional derivative with respect to the photon source; and, the vector potential obeys a differential equation in which the photon source is supplemented by a vectorial part of the electron-positron Green's function. (It looks better than it sounds.) It is remarked that, in addition to such one-particle Green's functions, one can also have multiparticle Green's functions.

The second note begins with:

> In all the work of the preceding note there has been no explicit reference to the particular states on (the space-like surfaces) that enter the definitions of the Green's functions. This information must be contained in boundary conditions that supplement the differential equations. We shall determine these boundary conditions for the Green's functions associated with vacuum states on both (surfaces).

And then:

> We thus encounter Green's functions that obey the temporal analog of the boundary condition characteristic of a source radiating into space. In keeping with this analogy, such Green's functions can be derived from a retarded proper time Green's function by a Fourier decomposition with respect to the mass.

The text continues with the introduction of auxiliary quantities:

> The mass operator M that gives a non-local extension to the electron mass; a somewhat analogous photon polarization operator P; and Γ, the non-local extension of the coupling between the electromagnetic field and the fields of the charged particles. Then, in the context of two-particle Green's functions, there is the interaction operator I.

The various operators that enter in the Green's function equations M, P, Γ, I, can be constructed by successive approximation. Perturbation theory, as applied in this manner, must not be confused with the expansion of the Green's functions in powers of the charge. The latter procedure is restricted to the treatment of scattering problems.

Then one reads:

It is necessary to recognize, however, that the mass operator, for example, can be largely represented in its effect by an alteration in the mass constant and by a scale change of the Green's function. Similarly, the major effect of the polarization operator is to multiply the photon Green's function by a factor, which everywhere appears associated with the charge. It is only after these renormalizations have been performed that we deal with wave equations that involve the empirical mass and charge, and are thus of immediate physical applicability.

In the period 1951–1952, two colleagues of mine at Harvard, and I, wrote a series of papers under the title "Electrodynamic Displacements of Atomic Energy Levels." The third paper, which does not carry my name, is subtitled "The Hyperfine Structure of Positronium." I quote a few lines:

The discussion of the bound states of the electron-positron system is based upon a rigorous functional differential equation for the Green's function of that system.
    And,
Theory and experiment are in agreement.

As for the rest of the 1950's, I focus on two highlights. First: although it could have appeared any time after 1951, it was 1958 when I published "The Euclidean Structure of Relativistic Field Theory." Here is how it begins:

The nature of physical experience is largely conditioned by the topology of space-time, with its indefinite Lorentz metric. It is somewhat remarkable, then, to find that a detailed correspondence can be established between relativistic quantum field theory and a mathematical image based on a four-dimensional Euclidean manifold. The objects that convey this correspondence are the Green's functions of quantum field theory, which contain all possible physical information. The Green's functions can be defined as vacuum-state expectation values of time-ordered field products.

I well recall the reception this received, running the gamut from "It's wrong" to "It's trivial." It is neither.

Second (highlight):

Another Harvard colleague and I had spent quite some time evolving the techniques before we published a 1959 paper entitled "Theory of Many-Particle Systems." It was intended to bring the full power of quantum field theory to bear on the problems encountered in solid state physics, for example. That required the extension of vacuum Green's functions, which refer to absolute zero temperature, into those for finite temperature. This is accomplished by a change of boundary conditions, which become statements of periodicity, or anti-periodicity, for the respective BE or FD statistics, in response to an imaginary time displacement.

As an off shoot of this paper, I published in 1960, "Field Theory of Unstable Particles." Here is how it begins:

> Some attention has been directed recently to the field theoretic description of unstable particles. Since this question is conceived as a basic problem for field theory, the responses have been some special device or definition, which need not do justice to the physical situation. If, however, one regards

the description of unstable particles to be fully contained in the framework of the general theory of Green's function, it is only necessary to emphasize the relevant structure of these functions. That is the purpose of this note. What is essentially the same question, the propagation of excitations in many-particle systems where stable or long-lived "particles" can occur under exceptional circumstances, has already been discussed along these lines.

One might be forgiven for assuming that this saga of George and me effectively ended with this paper. But that was 1/3 century ago!

To set the stage for what actually happened, I remind you that operator field theory is an extrapolation of ordinary quantum mechanics, with its finite number of degrees of freedom, to a continuum labeled by the spatial coordinates. The use of such space-time dependent variables presumes the availability, in principle, of unlimited amounts of momentum and energy. It is, therefore, a hypothesis about all possible phenomena of that type, the vast majority of which lies far outside the realm of accessible physics. In honor of a failed economic policy, I call such procedures: trickle-down theory.

In the real world of physics, progress comes from tentative excursions beyond the established framework of experiment and theory — the grass roots — indeed, the Green grass roots. What is sought here, in contrast with the speculative approach of trickle-down theory, is a phenomenological theory — a coherent account of the phenomena that is anabatic (from anabasis: going up).

The challenge was to reconstruct quantum field theory, without operator fields. The source concept was introduced in 1951 as a mathematical device — it was a source of fields. It took 15 years to appreciate that, with a finite, rather than an unlimited, supply of energy available, it made better sense to use the more physical-if idealized- concept of a particle source. Indeed, during that time period one had become accustomed to the fact that to study a particle of high energy physics, one had to create it. And, the act of detection involved the annihilation of that particle.

This idea first appeared in an article, entitled "Particles and Sources," which recorded a lecture of the 1966 *Tokyo Summer Lectures in Theoretical Physics*. The preface begins with:

> It is proposed that the phenomenological theory of particles be based on the source concept, which is abstracted from the physical possibility of creating or annihilating any particle in a suitable collision. The source representation displays both the momentum (energy) and the space-time characteristics of particle behavior.

Then, in the introduction, one reads:

> Any particle can be created in a collision, given suitable partners, before and after the impact to supply the appropriate values of the spin and other quantum numbers, together with enough energy to exceed the mass threshold. In identifying new particles it is basic experimental principle that the specific reaction is not otherwise relevant. Then, let us abstract from the physical presence of the additional particles involved in creating a given one (this is the vacuum) and consider them simply as the source of the physical properties that are carried by the created particle. The ability to give some localization in space and time to a creation act may be represented by a corresponding coordinate dependence of a mathematical source function, $S(x)$. The effectiveness of the source in supplying energy and momentum may be described by another mathematical source function, $S(p)$. The complementarity of these source aspects can be given its customary

quantum interpretation: $S(p)$ is the four-dimensional Fourier transformation of $S(x)$.

The basic physical act begins with the creation of a particle by a source, followed by the propagation (aha!) of that particle between the neighborhoods of emission and detection, and is closed by the source annihilation of the particle. Relativistic requirements largely constrain the structure of the propagation function — Green's function.

We now have a situation in which Green's function is not a secondary quantity, implied by a more fundamental aspect of the theory, but rather, is a primary part of the foundation of that theory. Of course fields, initially inferred as derivative concepts, are of great importance, as witnessed by the title I gave to the set of books I began to write in 1968: *Particles, Sources, and Fields*.

The quantum electrodynamics that began to emerge in 1947 still bothers some people because of the divergences that appear prior to renormalization. That objection is removed in the phenomenological source theory where there are no divergences, and no renormalization.

As another example of such clarification I cite a 1975 paper entitled " Casimir Effect in Source Theory." The abstract reads:

> The theory of the Casimir effect, including its temperature dependence is rederived by source theory methods, which do not employ the concept of (divergent) zero point energy. What source theory does have is a photon Green's function, which changes in response to the change of boundary conditions, as one conducting sheet is pushed into the proximity of another one.

A few years later, I and two colleagues at the University of California (UCLA), who had joined me from Harvard with their new doctorates, extended this treatment to dielectric bodies where forces of attraction also appear.

Having said this, I can move up to the present day, and the fascinating phenomenon of coherent sonoluminescense.

It has only recently been discovered that a single air bubble in water can be stabilized by an acoustical field. And, that the bubble emits pulses of light, including ultra violet light, in synchronism with the sonic frequency.

During the phase of negative acoustical pressure the bubble expands. That is followed by a contraction which, as Lord Rayleigh already recognized in his 1917 study of cavitation, turns into run away collapse. The recent measurements find speeds in excess of Mach 1 in air.

Then the collapse abruptly slows, and a blast of photons is emitted. In due time, the expansion slowly begins, and it all repeats, and repeats.

When confronted with a new phenomenon, everyone tends to see in it something that is already familiar. So, when told about this new aspect of sonoluminescence, I immediately said "It's the Casimir effect!" Not the static Casimir effect, of course, but the dynamical one of accelerated dielectric bodies. I have had no occasion to change my mind.

I can imagine a member of this audience thinking: "That's nice, but what is the role of George Green in this?"

Looking in at the center of the water container, one sees a steady blue light. A photomultiplier tube registers the succession of pulses, each containing a substantial number of photons, which can be an incomplete count because, deep in the ultraviolet, water becomes opaque.

A quantum mechanical description seeks the probabilities of emitting various numbers of photons, all of which probabilities are referred to the basic probability, that for emitting no photons. The latter probability dips below one — in some analogy with synchrotron radiation — because of the self-action carried by the electromagnetic field, as described by Green's function. And that function must obey the requirements imposed by an accelerated surface discontinuity, with water, the dielectric material, on one side, and a dielectric vacuum, air, on the other side. Carrying out that program is — as one television advertiser puts it — job one. Very fascinating, indeed.

So ends our rapid journey through 200 years. What, finally, shall we say about George Green? Why, that he is, in a manner of speaking, alive, well, and living among us.

PRELIMINARY REMARKS BEFORE BEGINNING HIS TECHNICAL TALK

Bryce DeWitt

University of Texas at Austin

Referring back to Abe's remarks I should first like to expand on Abe's reply to Gleick on the so-called "disadvantage" the Schwinger students were supposed to have found themselves at, in the late forties and early fifties. In the book by Sam Schweber that Abe mentions I am quoted as saying (this is off of a tape that I sent Sam ten years ago):

"He taught us to think in a more unconstrained way. It's perhaps paradoxical, but the limitations imposed by Schwinger's not mentioning Feynman or Dyson led us to adopt a broader perspective on physics. I think this is really true. It is true in my case."

Abe remarks that he and Bob Karplus were already mixing Schwinger techniques and Feynman graphs for an attack on the positronium problem. I did not absorb those techniques until much later. But no matter.

I was not given a thesis topic by Julian. I chose it myself. I had a strong feeling that Einstein's gravity theory was in a sort of limbo, detached from the rest of physics, and that it was a shame that such a beautiful theory should be so ignored. I proposed to drag it forcibly into the then modern world by redoing Schwinger's QED calculations with the gravitational field added. I was very naive in those days. Two years later, when I met Pauli and, hoping to go to Zürich, told him that I was trying to quantize the gravitational field, Pauli sat for several seconds, alternately nodding and shaking his head (as those who remember him know) and then said "That is a very important problem. But it will take somebody really smart."

Schwinger, on the other hand, accepted my proposal and said "Go ahead." Naturally I ran into difficulties. Schweber has recorded my saying that "I probably saw Schwinger, during my work on the thesis, a total of twenty minutes." I think that is about right. The main occasion I remember is sitting out in the hall beside Walter, who was also waiting to see the great man. Walter and I got to talking about Van Vleck whom I liked very much, and Walter remarked that he was not only a nice guy but a great physicist. I had just finished taking a group theory course from Van, taught in a very mediocre way, and I thought Walter was pulling my leg. As I say, I was very naive in those days.

When I got to see Schwinger he simply told me to drop the electron-positron field and concentrate on the interaction between gravitons and photons. This enabled me to get the thesis done. When it was all bound together as a book it was both mediocre and huge. When I handed it to Oppenheimer for his inspection on my arrival at the Institute for Advanced Study in 1949, he simply hefted it in his hand and, saying "monumental opus," handed it back to me.

I was able to finish the thesis because in modern parlance, I had worked on the photon propagator to only 1-loop order. But I was also aware that beyond 1-loop the waters were deep indeed. It was some years before I could see my way to dealing with quantum gravity to all orders, and in the meantime I dallied with canonical quantization, which does great violence to general covariance. During these years there was, however, one brief episode in which Schwinger played a very important role. Again I give you my remarks as recorded by Sam Schweber:

"I was a member of the Institute for Advanced Study in 1954 and was present when Schwinger came down and gave a marathon series of lectures, for a total of eleven hours — seven of them one day and four the next .... It was all extremely formal but extremely beautiful .... That marathon lecture had an enormous impact on me and has affected my research life ever since .... The lectures were never published. I have never seen a paper of Schwinger's that included the material he gave that day. They contained the effective action ..., the Legendre transform, the superdeterminant that is involved when you have simultaneously bosons and fermions."

The content of these lectures lay somewhat dormant in my mind, for the next three or four years. The missing ingredient was an understanding of Green's functions in curved spacetime. And this leads to a funny story: On my preliminary oral examination to qualify me to proceed to work on the Ph.D., Van Vleck asked me "What is a Green's function?" Since I didn't know, I said so. This so shocked the committee that Van came around to see me afterwards and suggested that perhaps I should consider going into experimental physics. To the delight of the gods, I am sure, the bulk of my subsequent life as a physicist has dealt with Green's functions, to which I now turn.

THE USES AND IMPLICATIONS OF CURVED-SPACETIME PROPAGATORS: A PERSONAL VIEW*

Bryce DeWitt

*Center for Relativity, Department of Physics
The University of Texas at Austin, Austin, Texas*

I am deeply honored by the award that has just been given to me, especially because I am from a generation of physicists sufficiently old to have known Paul Dirac personally and to have had numerous occasions to talk to him and to admire him.

No one was more surprised than I when the award was announced. I suppose what one is expected to do on an occasion like this is to reflect on one's past life. I find I have begun to do this with disturbing frequency of late. Such reflections are probably harmless enough if they do not become the occasion for boring one's fellows

---

*Based on a lecture given at the International Centre for Theoretical Physics, Trieste, on the occasion of receiving the Dirac Medal. Also in *Forty More Years of Ramifications: Spectral Asymptotics and Its Applications*, ed. by S. A. Fulling and F. J. Narcowich, **Discourses in Mathematics and Its Applications**, No. 1, © Department of Mathematics, Texas A&M University, College Station, Texas, 1991, pp. 27–48.

with anecdotes. I hope that the following outline of some of my early professional activity will not fall in this category.

The Falling-Charge Problem

I owe a considerable debt to my first student, Robert Brehme, who, in the late 1950's, asked me whether he could work on the problem: Does the equivalence principle apply to charged matter? I did not at first regard this as a very interesting problem. I had immediately translated it in my mind to the question: Does a falling charge radiate? And I saw no reason why it should not. In my view the issue of the equivalence principle was a red herring. The principle was never meant to apply other than locally to physical phenomena, and a charged particle is hardly a local object in view of the extended Coulomb field that it carries with it.

I was at that time trying to develop a canonical formalism for the gravitational field with the aim of creating a quantum theory of gravity, and I hoped that Brehme would assist me in this work. In fact, the work bogged down in the usual difficulties familiar to anyone who has tried to construct, and make sense of, a canonical quantum theory of gravity. So, in desperation, I agreed to let Brehme investigate the falling-charge problem; but I insisted that he do it properly. He was to begin by studying Dirac's famous 1938 paper on the classical radiating electron, in which all calculations are performed in a manifestly Lorentz covariant manner. He was then to translate this paper into the language of curved spacetime, keeping all the derivations manifestly generally covariant. He was not to introduce a special coordinate system at any stage.

The first obstacle he encountered was the problem of wave propagation in curved spacetime. Nobody seemed to have looked at this problem, at least in the physics literature. At length I discovered its solution, or at least part of its solution, in Hadamard's book *Lectures* on *Cauchy's Problem in Linear Partial Differential Equations* (Yale, 1923). In this book Hadamard does not use covariant notation or terminology, but it is easy to recast his results into covariant form. The prototype of all the equations he considers is that of a massless scalar field in a curved Lorentzian manifold of $1 + (n-1)$ dimensions:

$$\phi_{;\mu}{}^{\mu} = 0. \tag{1}$$

(Here a semicolon followed by p indices denotes a covariant derivative of pth order.) Hadamard introduces what he calls the "elementary solution" of this equation, based on an arbitrary fixed point z, and he notes important differences between the even and odd dimensional cases.

In a spacetime of $1 + 3$ dimensions ($n = 4$), the elementary solution takes the form

$$\phi = \frac{1}{8\pi^2}\left(\frac{u}{\sigma} + v\ln|\sigma| + w\right), \tag{2}$$

where σ is Synge's *world function* [1], equal to half the square of the geodetic distance between the fixed point z and the point x at which ϕ is evaluated, and u, v, w are certain smooth auxiliary functions. If one introduces the Van Vleck–Morette determinant [2,3]

$$D(x, z) = \det\left[-\frac{\partial^2}{\partial x^\mu \partial z^\alpha}\sigma(x, z)\right], \tag{3}$$

and its scalarized version

$$\Delta(x, z) = g^{-1/2}(x) D(x, z) g^{-1/2}(z), \quad g = |\det(g_{\mu\nu})|, \tag{4}$$

one finds, upon inserting (2) into (1), that

$$u = \Delta^{1/2} \tag{5}$$

(up to a normalization constant that has already been selected in Eq. (2)) and that v and w may be expressed as power series in σ:

$$v = \sum_{r=0}^{\infty} v_r \sigma^r, \quad w = \sum_{r=0}^{\infty} w_r \sigma^r. \tag{6}$$

The coefficients v_r and w_r are smooth functions of x and z, satisfying certain differential recursion relations.

It turns out that the recursion relations completely determine the v_r but not the w_r. The coefficient w_0 may be chosen arbitrarily. This corresponds to the possibility of adding to any particular solution of Eq. (1) an arbitrary smooth solution. The recursion relations for the v_r are simplified if one replaces them by other coefficients a_r related by

$$v_r = \frac{(-1)^{r+1}}{2^{r+1} r!} \Delta^{1/2} a_{r+1}. \tag{7}$$

One then finds, starting from $a_0 = 1$:

$$\sigma_{;}{}^\mu a_{r;\mu} + r a_r = \Delta^{-1/2} (\Delta^{1/2} a_{r-1})_{;\mu}{}^\mu, \quad r = 1, 2, 3, \ldots. \tag{8}$$

In obtaining these recursion relations, as well as the basic forms (2) and (5), one makes use of the following fundamental relations satisfied by the world function and the Van Vleck–Morette determinant:

$$\frac{1}{2} g^{\mu\nu}(x) \frac{\partial \sigma}{\partial x^\mu} \frac{\partial \sigma}{\partial x^\nu} = \frac{1}{2} g^{\alpha\beta}(z) \frac{\partial \sigma}{\partial z^\alpha} \frac{\partial \sigma}{\partial z^\beta} = \sigma, \tag{9}$$

$$\Delta^{-1}(\Delta \sigma_{;}{}^\mu)_{;\mu} = D^{-1}(D\sigma_{;}{}^\mu)_{;\mu} = n. \tag{10}$$

It should be stated that although the general structure of Hadamard's elementary solution, as exhibited by expression (2), is valid for all x and z, the expansions (6) are useful only when x is close to z and fail completely if there are conjugate points lying on the geodesic between x and z.

Green's Functions

Since expression (2) becomes singular when $\sigma = 0$, it cannot, without further investigation, strictly be regarded as solving Eq. (1) on the light cone through z. Careful analysis shows that it nevertheless does. Expression (2) also solves Eq. (1) in positive definite

(*Euclidean*) 4-dimensional manifolds *except* when $\sigma = 0$. For these manifolds σ is positive, vanishing only when $x = z$, and Hadamard's elementary solution becomes a Green's function. To get a Green's function in the Lorentzian case one must multiply the Hadamard solution by i and replace σ by $\sigma + i0$ in the singular terms, obtaining

$$G(x, z) = \frac{i}{8\pi^2} \left[\frac{\Delta^{1/2}}{\sigma + i0} + v \, ln(\sigma + i0) + w \right]. \tag{11}$$

With w appropriately chosen this is the so-called *Feynman propagator* for the massless scalar field in the given curved spacetime.

Since the differential operator acting on ϕ in Eq. (1) is real, the real part of any of its Green's functions is also a Green's function. Taking the real part of expression (11) and making use of the well known identities

$$ln(\sigma + i0) = ln|\sigma| + \pi i \theta(-\sigma), \tag{12}$$

$$\frac{1}{\sigma + i0} = \frac{d}{d\sigma} ln(\sigma + i0) = \frac{1}{\sigma} - \pi i \delta(\sigma), \tag{13}$$

where θ is the familiar step function, one obtains the Green's function

$$\bar{G}(x, z) = \frac{1}{8\pi} \left[\Delta^{1/2} \delta(\sigma) - v \theta(-\sigma) \right]. \tag{14}$$

Three important properties of this function may be noted: (1) It is independent of w and is hence unique. (2) It vanishes for spacelike separations of the points x and z, i.e., for $\sigma > 0$. (3) Although it has the same delta-function singularity on the light cone as has the flat-spacetime Green's function used by Dirac in his 1938 paper, it does not generally vanish *inside* the light cone. This last property is an expression of the fact that a plane or spherical sharp pulse of "radiation" satisfying Eq. (1) does not remain a sharp pulse, as it would in flat spacetime, but generally develops a "tail." Hadamard devotes a lot of attention to the so-called *tail function* $v(x, z)$ and asks the question: In what geometries does it vanish? He refers to the geometries for which $v \neq 0$ as those for which *Huyghens' principle* fails.

Brehme and I, too, devoted a lot of attention to the tail function, but for the quite different reason that it turns out to be solely responsible for the failure of the naive equivalence principle when applied to falling charges in empty space. To uncover this fact we had first to introduce retarded and advanced Green's functions and to generalize Eq. (1) to the case of vector fields. The Green's function (14) is actually the average of the retarded and advanced Green's functions, denoted here by G^- and G^+, respectively. The latter are obtained by chopping $\bar{G}$ with the *temporal step function*

$$\theta(x,z) \equiv \begin{cases} 1 & \text{if } x \text{ lies to the future of } z \\ 0 & \text{if } x \text{ lies to the past of } z \end{cases}, \tag{15}$$

the terms "past" and "future" in this definition being relative to any foliation of spacetime into spacelike hypersurfaces. Thus

$$\begin{aligned} G^-(x,z) &= 2\theta(x,z)\bar{G}(x,z), \\ G^+(x,z) &= 2\theta(z,x)\bar{G}(x,z). \end{aligned} \tag{16}$$

The generalization to the case of the electromagnetic field begins with

$$A_{\mu;\nu}{}^\nu - R_\mu{}^\nu A_\nu = 0, \tag{17}$$

which is the equation satisfied by the electromagnetic vector potential A_μ in empty space when the Lorentz gauge condition $A^\mu{}_{;\mu} = 0$ is imposed. One introduces an elementary solution similar to expression (2) and repeats the steps taken in the case of the scalar field equation (1), arriving at a Green's function $\bar{G}_{\mu\alpha}(x,z)$ having the form

$$\bar{G}_{\mu\alpha} = \frac{1}{8\pi}\left[\Delta^{1/2}\bar{g}_{\mu\alpha}\delta(\sigma) - v_{\mu\alpha}\theta(-\sigma)\right], \tag{18}$$

where the tail function $v_{\mu\alpha}$ again has an expansion valid when x is close to z:

$$v_{\mu\alpha}(x,z) = \sum_{r=0}^{\infty} v_{r\mu\alpha}\sigma^r. \tag{19}$$

The coefficients $v_{r\mu\alpha}$, like the coefficients v_r in Eq. (6), can be expressed in terms of alternative coefficients $a_{r\mu\alpha}$,

$$v_{r\mu\alpha} = \frac{(-1)^{r+1}}{2^{r+1}\,r!} \Delta^{1/2} a_{r+1\,\mu\alpha}, \qquad (20)$$

where the latter satisfy, starting from $a_{0\mu\alpha} = \bar{g}_{\mu\alpha}$, the differential recursion relations

$$\sigma_{;}^{\nu} a_{r\mu\alpha;\nu} + r a_{r\mu\alpha}$$
$$= \Delta^{-1/2} (\Delta^{1/2} a_{r-1\,\mu\alpha})_{;\nu}{}^{\nu} - R_\mu{}^\nu a_{r-1\,\nu\alpha}, \qquad r = 1,2,3\ldots. \qquad (21)$$

The chief new feature of the Green's function (18) is the appearance of the *geodetic parallel displacement matrix* $\bar{g}_{\mu\alpha}$ defined by

$$\sigma_{;}^{\nu} \bar{g}_{\mu\alpha;\nu} = 0, \quad \lim_{x\to z} \bar{g}_{\mu\alpha}(x,z) = g_{\mu\alpha}(z). \qquad (22)$$

The appearance of this matrix function is an expression of the fact that the polarization vector in any sharp pulse of electromagnetic radiation is always propagated in a parallel fashion along the null geodesic rays perpendicular to the pulse front.

The Equation of Motion

Although the Green's function (18), and the retarded and advanced Green's functions obtained from it by chopping, stem from Eq. (17) which is based on a particular choice of gauge, the retarded and advanced electromagnetic fields of a given charge-current source, calculated with the aid of these Green's functions, are independent of the gauge choice. In particular, the retarded and advanced Liénard–Wichert fields of a moving charge, computed with these Green's functions, are gauge independent. Precisely the latter fields are needed in generalizing Dirac's 1938 paper to the case of curved spacetime. After a long and tedious calculation [4] of the mechanical and electromagnetic energy balance across a cylindrical 3-surface surrounding the world line of a spinless particle of charge e and experimental mass m moving in an external electromagnetic field $F_{\alpha\beta}$, in which coincidence limits $(x \to z)$ of the functions $\bar{g}_{\mu\alpha}, v_{\mu\alpha}$ and their derivatives are computed with the aid of the recursion relations (21) and covariant derivatives of Eqs. (9), (10), and (22), one finds the following equation of motion for the particle:

$$m\ddot{z}^\alpha = eF^\alpha{}_\beta \dot{z}^\beta + \frac{2}{3}e^2(\dddot{z}^\alpha - \dot{z}^\alpha \ddot{z}^\gamma)$$

$$- \frac{1}{3}e^2(R^\alpha{}_\beta + \dot{z}^\alpha \dot{z}^\gamma R_{\alpha\beta})\dot{z}^\beta + \frac{2}{3}e^2 \dot{z}^\beta \int_{-\infty}^{\tau} f^\alpha{}_{\beta\gamma'} \dot{z}^{\gamma'}(\tau')\,d\tau' , \quad (23)$$

where $f^\alpha{}_{\beta\gamma'}$ is the curl of the tail function:

$$f_{\alpha\beta\gamma'}(z,z') = v_{\beta\gamma';\alpha}(z,z') - v_{\alpha\gamma';\beta}(z,z') , \quad (24)$$

and the dots denote covariant differentiation with respect to the proper time τ of the particle.

In empty space $F_{\alpha\beta} = 0$ and $R_{\alpha\beta} = 0$, and the first and third terms on the right of Eq. (23) vanish. One sees in this case that the only thing preventing the charged particle from executing geodetic motion (and hence obeying the naive equivalence principle) is the integral of the curl of the tail function over the past history of the particle. When the gravitational field is weak (small curvature), this *tail integral* can be accurately computed, not by means of the expansion (19) but with the aid of first order perturbation theory. Moreover, a quasi-Minkowskian coordinate system can be introduced, and if the motion with respect to this coordinate system is slow one finds [5] that Eq. (23) in empty space reduces effectively to

$$m(\ddot{\mathbf{r}} + \nabla\phi) = -\nabla\psi - \frac{2}{3}e^2 \dot{\mathbf{r}} \cdot \nabla\nabla\phi , \quad (z^\mu) = (t, \mathbf{r}) , \quad (25)$$

where the dots now denote differentiation with respect to ordinary time t and where

$$\phi(\mathbf{r}) = -G \int \frac{\rho(\mathbf{r}')}{|\mathbf{r}-\mathbf{r}'|} d^3\mathbf{r}', \qquad (26)$$

$$\psi(\mathbf{r}) = \frac{1}{2} e^2 G \int \frac{\rho(\mathbf{r}')}{|\mathbf{r}-\mathbf{r}'|^2} d^3\mathbf{r}', \qquad (27)$$

$\rho(\mathbf{r}')$ being the density of the mass producing the gravitational field. G is the gravity constant, ϕ is the usual Newtonian potential, and ψ is an anomalous potential that makes a retrograde contribution to perihelion precession in the case of orbital motion.

The right hand side of Eq. (25) consists of two parts: a conservative force described by the potential ψ and a nonconservative force of the frictional type, depending linearly on the velocity. The conservative force can be shown [5] to arise from the fact that the (passive) gravitational mass of the particle is not concentrated at a point but is partly distributed as electric field energy in the space surrounding the particle. The nonconservative force arises from a back-scatter process in which the Coulomb field of the particle, as it sweeps over the "bumps" in spacetime, receives "jolts" that are propagated back to the particle. The direct role of curvature can be seen by noting that the ith component of the nonconservative force in the weak-field slow-motion limit is equal to $\frac{2}{3} e^2 R_{i0j0} \dot{z}^j$ where $R_{\mu\nu\sigma\tau}$ is the curvature tensor.

The back-scatter process, which also produces radiation that escapes to infinity, originates well outside the classical radius of the particle. Dividing Eq. (25) by m, one sees that the terms on the right are smaller than those on the left by the ratio of the classical radius e^2/m to the size of the mass distribution producing the gravitational field. Using $\ddot{\mathbf{r}} = -\nabla\phi$ as a first approximation to the motion, one also sees that the nonconservative force can be written in the form

$$-\frac{2}{3} e^2 \frac{d}{dt} \nabla\phi \approx \frac{2}{3} e^2 \dddot{\mathbf{r}}, \qquad (28)$$

which is like the radiative damping force associated with accelerations produced by nongravitational forces. The similarity must not be pushed too far, however, since damping in the present case is really of the frictional type and cannot lead to runaway behavior.

Reciprocity Relations

My debt to Robert Brehme is not limited to the fact that his thesis work led to my learning these lovely results. I have also learned other lovely results by reflecting on the general properties of Green's functions, not only for boson fields but for fermion fields as well. Let F be the differential operator that describes the propagation of a small (strictly infinitesimal) disturbance ψ on a dynamical background field. The equations satisfied by the background field may be linear or nonlinear. In either case the operator F is linear, and ψ is known as a *Jacobi field*. If the dynamical equations of the background field derive from an action principle, then F is self-adjoint, being symmetric and real for real boson fields and antisymmetric and imaginary for real fermion fields. Moreover, its retarded and advanced Green's functions $G^{\pm}(x, x')$ satisfy the following reciprocity relations:

$$G^{\pm}(x, x') = \left\{ \begin{array}{ll} G^{\mp T}(x', x) & \text{for boson fields} \\ -G^{\mp T}(x', x) & \text{for fermion fields} \end{array} \right\}. \qquad (29)$$

(Here the field indices have been suppressed and the superscript "T" denotes the matrix transpose.) These relations have the following corollary:

$$\tilde{G}(x, x') = \left\{ \begin{array}{ll} -\tilde{G}^{T}(x', x) & \text{for boson fields} \\ \tilde{G}^{T}(x', x) & \text{for fermion fields} \end{array} \right\}, \qquad (30)$$

where

$$\tilde{G}(x, x') \equiv G^{+}(x, x') - G^{-}(x, x'). \qquad (31)$$

The function $\tilde{G}$ is important because it solves the Cauchy problem for the equation

$$F\psi = 0. \qquad (32)$$

Thus let Σ be an arbitrary spacelike hypersurface. Then solutions of Eq. (32) are given in terms of Cauchy data on Σ by the integral

$$\psi(x) = \int_\Sigma G(x,x')\overleftrightarrow{W}^{\mu'}\psi(x')\,d\Sigma_{\mu'}, \qquad (33)$$

where $\overleftrightarrow{W}^{\mu'}$ is the *Wronskian operator* associated with F, defined by

$$\int_\Omega [\psi_1^\dagger(F\psi_2) - (F^\dagger\psi_1)^\dagger\psi_2]\,d^n x = \int_{\partial\Omega} \psi_1^\dagger\overleftrightarrow{W}^\mu\psi_2\,d\Sigma_\mu. \qquad (34)$$

The reciprocity relations were of interest to me because I had encountered them at about the same time (1960) in an entirely different context. I was making a great effort to study the famous but difficult paper by Bohr and Rosenfeld [6] on the measurability of the quantized electromagnetic field, with the aim of settling the analogous question whether there can be any measurement-theoretical meaning to quantizing the gravitational field. The chief issue concerns the effect of disturbances caused by the measurement process itself. A disturbance can be idealized as a change in the action functional S of a system:

$$S \to S + \varepsilon A. \qquad (35)$$

The quantity A is determined by the nature of the disturbance, and the coefficient ε describes its strength. To minimize the disturbance one considers the limit $\varepsilon \to 0$. The (retarded) change in any observable B, due to the change εA in the action, is then given by $\varepsilon D_A^- B$ where

$$D_A^- B = \int d^n x \int d^n x' B \frac{\overleftarrow{\delta}}{\delta\varphi(x)} G^-(x,x') \frac{\overrightarrow{\delta}}{\delta\varphi(x')} A, \qquad (36)$$

the field variables of the system being denoted by $\varphi(x)$. Replacing the retarded Green's function on the right of Eq. (36) by the advanced Green's function and defining an analogous quantity $D_A^+ B$, one easily sees that the reciprocity relations (29) imply

$$D_A^+ B = \begin{cases} -D_B^- A & \text{if both } A \text{ and } B \text{ are fermionic} \\ D_B^- A & \text{otherwise} \end{cases}, \qquad (37)$$

which may be stated in words thus: *The advanced effect of A on B is equal (up to a possible sign) to the retarded effect of B on A.* This result, with the words "advanced" and "retarded" omitted, was known to Lagrange [7] in his study of the mutual, instantaneously propagated, Newtonian perturbations of the planets by each other.

The Bohr–Rosenfeld Analysis and the Peierls Bracket

In the study made by Bohr and Rosenfeld, two dynamical entities were under consideration: a *system* and an *apparatus*. By an appropriate coupling between system and apparatus, the apparatus is able to record the value of some system observable A. If the action functionals of system and apparatus are denoted by S and Σ, respectively, the measurement may be described by a change in the total action of the form

$$S + \Sigma \to S + \Sigma + g\mathcal{X}, \tag{38}$$

where $g\mathcal{X}$ describes the coupling, g being an adjustable "coupling constant." The disturbance in the apparatus, produced by this coupling, is what constitutes the measurement. Of course, the system also gets disturbed by the coupling, and one tries to minimize this. Bohr and Rosenfeld found that the analysis of the latter disturbance requires one to work to second order in g and that to achieve maximum accuracy one must introduce a *compensation term* of this order into the coupling between system and apparatus. In the notation used here this corresponds to changing (38) to

$$S + \Sigma \to S + \Sigma + g\mathcal{X} - \frac{1}{2}g^2 D^-_{\mathcal{X}} \mathcal{X} \tag{39}$$

(see Ref. 8).

The coupling functional $\mathcal{X}$ must obviously depend on the dynamical variables of both system and apparatus. Let Π be the apparatus variable chosen to "store" the results of the measurement. If Π is to provide a permanent record, it must refer to a time interval to the future of the coupling interval on which $\mathcal{X}$ depends, and hence

$$D_\Pi^- \mathcal{X} = 0. \tag{40}$$

On the other hand, to produce a (retarded) change in Π given by

$$\delta \Pi = gA, \tag{41}$$

and hence effectively to *measure* A, one must choose $\mathcal{X}$ so that

$$D_{\mathcal{X}}^- \Pi = A. \tag{42}$$

Equation (41) is then obviously correct to first order. If the coupling is of the modified form given by (39) and if the apparatus is chosen massive compared to the system so that, in expansions of the disturbance equations, terms involving Green's functions for the apparatus may be neglected in comparison with analogous terms involving Green's functions for the system, then to second order in g, one finds [8]

$$\delta\Pi = gD_{\mathcal{X}}^-\Pi + g^2 D_{\mathcal{X}}^- D_{\mathcal{X}}^-\Pi - \frac{1}{2}g^2 D_{D_{\mathcal{X}}^-\mathcal{X}}^- \Pi$$

$$= gA + g^2 D_{\mathcal{X}}^- A - \frac{1}{2}g^2 D_{\mathcal{X}}^- A - \frac{1}{2}g^2 D_A^- \mathcal{X}. \tag{43}$$

If one additionally chooses $\mathcal{X}$ so that

$$D_{\mathcal{X}}^- A = D_A^- \mathcal{X} \tag{44}$$

(for example, by setting $\mathcal{X} = AX$ where X is an apparatus observable satisfying $D_X^- \Pi = 1$) then Eq. (43) reduces to (41) to *second* order. This has the consequence that, for small g, the accuracy in the measurement of A is limited only by the *a priori* sharpness of definition of the apparatus observable Π.

Such precision is generally lost when the attempt is made to measure two system observables A and B simultaneously. In this case one chooses two apparatus observables, Π_A and Π_B, to record the measurements, and a coupling of the form

$$g(\mathcal{X} + \mathcal{Y}) - \frac{1}{2}g^2 D_{(\mathcal{X}+\mathcal{Y})}^-(\mathcal{X} + \mathcal{Y}), \tag{45}$$

satisfying

$$D_{\mathcal{X}}^{-}\Pi_A = A, \quad D_{\mathcal{X}}^{-}\Pi_B = 0, \quad D_{\mathcal{X}}^{-}A = D_A^{-}\mathcal{X}, \qquad (46)$$

$$D_{\mathcal{Y}}^{-}\Pi_A = 0, \quad D_{\mathcal{Y}}^{-}\Pi_B = B, \quad D_{\mathcal{Y}}^{-}B = D_B^{-}\mathcal{Y}, \qquad (47)$$

$$D_{\Pi_A}^{-}\mathcal{X} = 0, \quad D_{\Pi_A}^{-}\mathcal{Y} = 0, \quad D_{\Pi_B}^{-}\mathcal{X} = 0, \quad D_{\Pi_B}^{-}\mathcal{Y} = 0. \qquad (48)$$

In addition, the apparatus observables must be independent, satisfying

$$D_{\Pi_A}^{-}\Pi_B = D_{\Pi_B}^{-}\Pi_A = 0. \qquad (49)$$

To second order the changes in these observables as a result of the coupling are given by

$$\delta\Pi_A = gD_{\mathcal{X}+\mathcal{Y}}^{-}\Pi_A + g^2 D_{\mathcal{X}+\mathcal{Y}}^{-} D_{\mathcal{X}+\mathcal{Y}}^{-}\Pi_A - \frac{1}{2}g^2 D_{D_{\mathcal{X}+\mathcal{Y}}^{-}(\mathcal{X}+\mathcal{Y})}^{-}\Pi_A$$

$$= gA + \frac{1}{2}g^2 D_{\mathcal{X}+\mathcal{Y}}^{-} A - \frac{1}{2}g^2 D_A^{-}(\mathcal{X}+\mathcal{Y})$$

$$= gA + \frac{1}{2}g^2(D_{\mathcal{Y}}^{-}A - D_A^{-}\mathcal{Y}) \qquad (50)$$

and, similarly,

$$\delta\Pi_B = gB + \frac{1}{2}g^2(D_{\mathcal{X}}^{-}B - D_B^{-}\mathcal{X}). \qquad (51)$$

The quantities appearing in parentheses on the right of these equations are actually Poisson brackets and were first presented in this form by Peierls [9]. The remarkable thing about Peierls' brackets is that they do not depend for their definition on the introduction of a canonical formalism. They are completely determined by the laws of propagation of Jacobi fields, and their definition emphasizes a global spacetime view of the dynamics. When I first realized that Bohr and Rosenfeld were dealing with Peierls brackets, I became quite excited, for reasons that I shall discuss in a moment. Let me first give the modern definition of the Peierls bracket of two observables A and B:

$$(A,B) \equiv \begin{cases} -D_A^- B - D_B^- A & \text{if both } A \text{ and } B \text{ are fermionic} \\ D_A^- B - D_B^- A & \text{otherwise} \end{cases}. \tag{52}$$

In view of the reciprocity relation (37), this definition can be recast in the form

$$(A,B) = D_B^+ A - D_B^- A = \int d^n x \int d^n x' A \frac{\overleftarrow{\delta}}{\delta\varphi(x)} \widetilde{G}(x,x') \frac{\overrightarrow{\delta}}{\delta\varphi(x')} B, \tag{53}$$

in which the dynamical field variables $\varphi(x)$ are reintroduced in the final expression. The Peierls bracket is seen to have the same symmetry as the supercommutator bracket in the quantum theory, the latter being defined by

$$[A,B] \equiv \begin{cases} AB + BA & \text{if both } A \text{ and } B \text{ are fermionic} \\ AB - BA & \text{otherwise} \end{cases}. \tag{54}$$

This allows one to introduce the heuristic (i.e., up to internal factor-ordering ambiguities) quantization rule

$$[A,B] = i(A,B), \quad \hbar = 1, \tag{55}$$

and to call the function $\widetilde{G}$ the *supercommutator function*. The Peierls' bracket can be shown [8,10] to satisfy the super Jacobi identity.

Quantum Mechanical Limitations on Measurement Accuracy

Since the quantities appearing in the analysis of measurement couplings are bosonic, one can now rewrite Eqs. (50) and (51) in the forms

$$\delta\Pi_A = gA + \frac{1}{2}g^2(\mathcal{Y},A), \quad \delta\Pi_B = gB + \frac{1}{2}g^2(\mathcal{X},B), \tag{56}$$

which may be solved for the "experimental values" of A and B:

$$A = \frac{\delta\Pi_A}{g} - \frac{1}{2}g(\mathcal{Y},A), \quad B = \frac{\delta\Pi_B}{g} - \frac{1}{2}g(\mathcal{X},B). \tag{57}$$

These equations yield the accuracy estimates

$$(\Delta A)^2 = \frac{(\Delta \Pi_A)^2}{g^2} + \frac{1}{4}g^2[\Delta(\mathcal{Y}, A)]^2, \tag{58}$$

$$(\Delta B)^2 = \frac{(\Delta \Pi_B)^2}{g^2} + \frac{1}{4}g^2[\Delta(\mathcal{X}, B)]^2. \tag{59}$$

Multiplying Eqs. (58) and (59) together, one finds that the minimum value of the product of the uncertainties ΔA and ΔB is

$$\Delta A \Delta B = \frac{1}{2}[\Delta \Pi_A \Delta(\mathcal{X}, B) + \Delta \Pi_B \Delta(\mathcal{Y}, A)], \tag{60}$$

this minimum occurring when $g^4 = 4\Delta A \Delta B / \Delta(\mathcal{Y}, A)\Delta(\mathcal{X}, B)$.

Under the best imaginable circumstances, the brackets $(\mathcal{Y}, A)$ and $(\mathcal{X}, B)$ will depend only weakly on the system dynamics, so that the uncertainties $\Delta(\mathcal{Y}, A)$ and $\Delta(\mathcal{X}, B)$ will arise primarily from the *a priori* imprecision of the apparatus variables. If we assume that the laws of quantum mechanics apply to the apparatus as well as the system, then we must have, in the relevant quantum state,

$$\Delta \Pi_A \, \Delta(\mathcal{X}, B) \geq \tfrac{1}{2}|\langle[\Pi_A, (\mathcal{X}, B)]\rangle| = \tfrac{1}{2}|(\Pi_A, (\mathcal{X}, B))|, \tag{61}$$

$$\Delta \Pi_B \, \Delta(\mathcal{Y}, A) \geq \tfrac{1}{2}|\langle[\Pi_B, (\mathcal{Y}, A)]\rangle| = \tfrac{1}{2}|(\Pi_B, (\mathcal{Y}, A))|. \tag{62}$$

But

$$(\Pi_A, (\mathcal{X}, B)) = -(\mathcal{X}, (B, \Pi_A)) - (B, (\Pi_A, \mathcal{X}))$$

$$= (B, D_{\mathcal{X}}^- \Pi_A) = (B, A), \tag{63}$$

$$(\Pi_B, (\mathcal{Y}, A)) = -(\mathcal{Y}, (A, \Pi_B)) - (A, (\Pi_B, \mathcal{Y}))$$

$$= (A, D_{\mathcal{Y}}^- \Pi_B) = (A, B), \tag{64}$$

whence it follows that

$$\Delta A \Delta B \geq \tfrac{1}{2}|(A, B)| = \tfrac{1}{2}|\langle[A, B]\rangle|. \tag{65}$$

This result, which Bohr and Rosenfeld obtained for the special case of quantum electrodynamics, says that measurements can, in principle, always be made to an accuracy equal to *but no better than* that allowed by the *a priori* uncertainties implied by the quantum mechanical formalism. After thoroughly absorbing the Bohr–Rosenfeld paper, I was able to show [11], by introducing idealized measuring instruments akin to gravitational wave antennae, that simultaneous measurements of spacetime averages of any pair of components of the Riemann tensor (projected onto a physical coordinate system consisting of an elastic medium carrying a field of clocks) can be made to an accuracy exactly equal to that allowed by the quantum formalism. Thus, quantizing the gravitational field is exactly as meaningful as quantizing the electromagnetic field. There is just one new limitation, which does not exist in the electromagnetic case: The sizes of the spacetime averaging domains must be large compared to the Planck length, 10^{-33} cm. Many arguments lead to the conclusion that standard notions of space and time, and even of probability itself, cease to have operational meaning below this scale. This is the domain in which string theory is supposed to have something new to say.

My excitement over the discovery of the role of the Peierls bracket in the Bohr–Rosenfeld analysis stemmed from the fact that Bohr and Rosenfeld, and also myself in the gravitational case, were able to confine our attention exclusively to observable, and hence *gauge invariant*, quantities. This meant that the Peierls bracket need never be defined for other than gauge invariant quantities. Indeed one can show [10] that as long as $\mathcal{A}$ and $\mathcal{B}$ in the double integral in Eq. (53) are gauge invariant, any choice of gauge conditions may be used to define the supercommutator function $\widetilde{G}$. All choices lead to the same bracket. This made it appear to me that by proceeding directly from the Peierls bracket, one could construct quantum theories of gauge systems — in particular, non-Abelian gauge systems — without having to introduce the tiresome canonical complications of constrained

Hamiltonian mechanics and bigger-than-physical Hilbert spaces. Indeed one *can* do this, and my efforts from 1962 to 1967, when my trilogy [12] on quantum gravity was published, were devoted to precisely this end. There is not space here to describe this effort, which used over and over again the properties of covariant Green's functions, and their determinants, for both basic fields and ghost fields and which finally culminated in the first derivation of the Feynman rules for non-Abelian gauge fields to all orders in arbitrary gauges. Little did I know in 1959 that Robert Brehme's problem would lead to all this.

The Schwinger Variational Principle

Before turning to more modern developments, I cannot resist describing one beautiful application of the Peierls bracket that is of fundamental practical importance. Let $|\text{in}\rangle$ and $|\text{out}\rangle$ denote state vectors referring to dynamical situations, respectively, in the past and future of the era of interest in any dynamical study. We may call this dynamically interesting era the *present*. Suppose the dynamics of the system are changed by introducing a change δS in the action fundamental, and suppose the dynamical variables out of which δS is built are taken from the *present*. $|\text{in}\rangle$ and $|\text{out}\rangle$ will typically be eigenvectors of sets of operators, denoted collectively by A and B, respectively, which are built out of dynamical variables taken from the past and future, respectively. This implies

$$D^-_{\delta S} A = 0, \quad D^-_B \delta S = 0. \tag{66}$$

Suppose we describe the changed dynamics of the system in terms of retarded boundary conditions. Then A will remain unaffected, but B will suffer the following change:

$$\delta B = D^-_{\delta S} B = D^-_{\delta S} B - D^-_B \delta S = (\delta S, B) = -i[\delta S, B], \tag{67}$$

in which Eqs. (55) and (66) have been used. An alternative statement, correct to first infinitesimal order, is

$$B + \delta B = (1 - i\delta S)B(1 + i\delta S), \tag{68}$$

which implies that, although the vector $|\text{in}\rangle$ remains unchanged by the changed dynamics, the vector $|\text{out}\rangle$ gets replaced by

$$|\text{out}\rangle + \delta|\text{out}\rangle = (1 - i\delta S)|\text{out}\rangle, \tag{69}$$

whence

$$\delta|\text{in}\rangle = 0, \quad \delta|\text{out}\rangle = -i\delta S|\text{out}\rangle, \tag{70}$$

$$\delta\langle\text{out}|\text{in}\rangle = i\langle\text{out}|\delta S|\text{in}\rangle. \tag{71}$$

The same result is also obtained with any other boundary conditions.

Equation (71) is *Schwinger's variational principle*. By letting δS be the result of varying external sources that have been introduced into the action functional and by introducing the notion of *functional Fourier transforms*, one can show [10] that the Schwinger principle leads immediately to the Feynman functional integral for the amplitude $\langle\text{out}|\text{in}\rangle$ and to all the practical applications that go with it. To be sure, the derivation of the Schwinger principle from the Peierls bracket is only heuristic because of the factor ordering ambiguities that the latter ignores, but the modern view is that the unitary transformation (68) is so beautiful that it should be taken as gospel and that the Schwinger–Feynman theory that follows from it should in the end solve all these trivialities.

The Feynman Propagator

In most practical applications the states represented by the vectors $|\text{in}\rangle$ and $|\text{out}\rangle$ are either vacuum states, "relative vacuum" states [10], scattering states, or coherent states, and the relevant Green's function appearing in perturbative expansions of the Feynman functional integral is the so-called *Feynman propagator*. The Feynman propagator is colloquially described as that Green's function (of the small-disturbance operator F) which propagates negative frequencies into the past and positive frequencies into the future. Strictly speaking, the terms "positive" and "negative frequencies" have meaning only with respect to a timelike Killing vector field. One of the important lessons learned in the last fifteen years, in the effort to extend quantum field theory to curved spacetime, is that the notions of *vacuum state* and *Fock space* are *relative*. They have to be generalized from the standard Minkowski vacuum of flat spacetime and defined with care, *relative* to stationary background fields. Once this is done one can show [10] that the usual definition of the Feynman propagator, in terms of the choice of an integration contour around poles in energy space, is equivalent, in the case of boson fields, to the following formal definition:

$$G = -\frac{1}{F + i0}. \tag{72}$$

(A slightly modified definition must be adopted in the fermion case [10].)

If the background field is not stationary, i.e., if a global timelike Killing vector field does not exist, one usually assumes the existence of stationarity in one or more asymptotic regions. The representation (72) is then still valid, for it implies the variational formula

$$\delta G = G \delta F G, \tag{73}$$

which maintains the desired boundary conditions in the asymptotic regions.

As an example of the utility of the representation (72), consider a scalar field propagating in external Yang–Mills and gravitational fields. The small-disturbance operator has the general form

$$F = -\pi_\mu g^{1/2} g^{\mu\nu} \pi_\nu - g^{1/2}(m^2 + V), \tag{74}$$

$$\pi_\mu = -i\frac{\partial}{\partial x^\mu} + G_\alpha A^\alpha{}_\mu, \tag{75}$$

$$[G_\alpha, G_\beta] = G_\nu c^\gamma{}_{\alpha\beta}, \tag{76}$$

where $A^\alpha{}_\mu$ are the Yang–Mills potentials, the $c^\gamma{}_{\alpha\beta}$ are the structure constants of the Yang–Mills group and the G_α are the generators of the representation of this group to which the scalar field belongs. The "potential" V may contain terms arising from self-coupling of the scalar fields, as well as a term proportional to the curvature scalar. The operator (74), although describing scalar disturbances, is actually general enough to serve as a prototype of all field-theoretical small-disturbance operators, for fields of nonvanishing spin may be regarded formally as scalar fields bearing extra indices referring to the local Lorentz group, with the spin-connection replacing $A^\alpha{}_\mu$. Moreover, the potential V need not be a multiple of the unit matrix but may be a more general matrix.

The Heat Kernel and Its Uses

When spacetime is curved, it is convenient to replace (72) by

$$g^{1/4} G g^{1/4} = -\frac{1}{g^{-1/4} F g^{-1/4} + i0} = i \int_0^\infty e^{ig^{-1/4} F g^{-1/4} s}\, ds, \quad (77)$$

in which the "$i0$" rule is replaced by an equivalent definition in terms of a complex Laplace transform. The insertion of the factors $g^{-1/4}$ is needed when the Laplace-transform definition is used because only $g^{-1/4} F g^{-1/4}$, not F, can be covariantly exponentiated. Introducing the more explicit coordinate representation, one now sees that (77) can be re-expressed in the form

$$g^{1/4}(x) G(x, x') g^{1/4}(x') = i \int_0^\infty K(x, x', s)\, ds, \quad (78)$$

where $K(x, x', s)$ describes the cumulative effect of the exponentiated operator $ig^{-1/4} F g^{-1/4} s$ and satisfies the differential equation

$$i \frac{\partial}{\partial s} K(x, x', s) = -g^{-1/4} F g^{-1/4} K(x, x', s)$$

$$= (-D_\mu D^\mu + m^2 + V) K(x, x', s), \quad (79)$$

together with the boundary condition

$$K(x, x', 0) = \delta(x, x'). \tag{80}$$

If spacetime were positive definite and if the variable s were rotated through 90° in the complex plane, Eq. (79) would be a kind of *heat equation* (actually a covariant diffusion equation, D_μ here denoting the covariant derivative and $D_\mu D^\mu$ the covariant Laplacian). It is for this reason that $K(x, x', s)$ is often called the *heat kernel*. Although Eqs. (78) and (79) have been known in the physics literature for at least three decades, mathematicians were independently making use of them during the same period for reasons having to do with differential topology and index theorems. Since the mathematicians were generally dealing with positive definite manifolds, they of course worked with the ordinary Laplace transform and the ordinary heat equation from the outset.

Starting from the known form of the heat kernel in flat, n-dimensional, empty spacetime, one can introduce the following Ansatz for it in the general case:

$$K(x, x', s)$$
$$= i(4\pi i s)^{-n/2} D^{1/2}(x, x') e^{(i/2s)\sigma(x,x') - im^2 s} \sum_{r=0}^{\infty} a_r(x, x')(is)^r. \tag{81}$$

Inserting this expression into Eq. (79) and making use of Eqs. (9) and (10), one obtains recursion relations for the a_r, starting from $a_0 = 1$, which show them to be identical with the coefficients already encountered in connection with the Hadamard elementary solution. When $m \neq 0$ one can actually insert the expansion (81) into Eq. (78) and obtain a definite expression for the function w in Eq. (11). (Remember that this function remained undetermined by Hadamard.) But one also discovers that (81) is only an asymptotic expansion. It fails to yield an accurate approximation to the Green's function when m is small. It also fails to yield a correct expression for the heat kernel when there is more than one geodesic between x and x'. It nevertheless turns out to be useful both in renormalization theory and in the derivation of anomalies.

To illustrate both of these uses it suffices to consider the WKB or "one-loop" approximation to the Feynman functional integral for the in-out amplitude. This is given by

$$\langle \text{out}|\text{in}\rangle \equiv e^{iW} \approx \text{const.} \times \left(\frac{\det G}{\det G^+}\right)^{1/2} e^{iS}. \tag{82}$$

The ratio of the two formal determinants in the last expression is the field-theoretic version of the Van Vleck–Morette determinant [2,3] of ordinary WKB theory. The denominator $(\det G^+)^{1/2}$ can be shown [10] to be a guarantor of unitarity and to permit the usual Wick rotation in the evaluation of integrals for Feynman graphs. It is also frequently ignored (as are the arcs at infinity in Wick rotations).

If one functionally differentiates the exponent W in Eq. (82) with respect to the background fields, one gets "in-out averages" of the corresponding sources:

$$\langle j_\alpha^\mu \rangle = \frac{\langle \text{out}|j_\alpha^\mu|\text{in}\rangle}{\langle \text{out}|\text{in}\rangle} = \frac{\delta W}{\delta A^\alpha{}_\mu}, \tag{83}$$

$$\langle T^{\mu\nu} \rangle = \frac{\langle \text{out}|T^{\mu\nu}|\text{in}\rangle}{\langle \text{out}|\text{in}\rangle} = 2\frac{\delta W}{\delta g_{\mu\nu}}. \tag{84}$$

(These equations are easy consequences of the Schwinger variational principle.) Now note that variations of the background fields produce variations in F. Taking the logarithm of Eq. (82) and varying it, one finds

$$\delta(W - S) = -\frac{1}{2}i \, \text{Tr}\left[(G - G^+)\delta F\right]$$

$$= -\frac{1}{2}i \, \text{Tr}\left[g^{1/4}(G - G^+)g^{1/4}\delta(g^{-1/4}Fg^{-1/4})\right]. \tag{85}$$

The trace Tr includes a bringing into coincidence of two spacetime points and an integration over spacetime. If it is understood that the points are to be brought together from spacelike directions (or after a Wick rotation), then the Green's function G^+ can be omitted and Eq. (85) replaced by

$$\delta(W - S) = -\frac{1}{2}i \ \text{Tr}\left[g^{1/4}Gg^{1/4}\delta(g^{-1/4}Fg^{-1/4})\right]$$

$$= \frac{1}{2} \ \text{Tr} \int_0^\infty e^{ig^{-1/4}Fg^{-1/4}s}\delta(g^{-1/4}Fg^{-1/4})\,ds$$

$$= \delta\left(\frac{1}{2} \ \text{Tr} \int_0^\infty e^{ig^{-1/4}Fg^{-1/4}s}\frac{1}{is}\,ds\right), \qquad (86)$$

whence it follows that

$$W - S = \int w(x)\,d^n x + \text{const.}, \qquad (87)$$

where

$$w(x) = \frac{1}{2}\int_0^\infty \text{tr } K(x,x,s)\frac{1}{is}\,ds. \qquad (88)$$

Zeta Function Regularization

The integral (88) is actually divergent at the lower limit. In their study of topological questions and index theorems, mathematicians have introduced a useful trick to give the integral meaning. They first introduce a generalized zeta function, defined by

$$\zeta(x,z) \equiv \frac{1}{\Gamma(z)}\int_0^\infty (i\mu^2 s)^{z-1} K(x,x,s)\,i\mu^s\,ds, \qquad (89)$$

in the region of the complex z plane where the integral is convergent, and defined elsewhere by analytic continuation. Here μ is an arbitrary mass parameter introduced to yield the dimensionless combination $i\mu^2 s$ so that the analytic continuation can be consistently carried out.

One can then use expression (81), which *does* represent the singularity structure of the heat kernel correctly as $s \to 0$, to infer that the analytic continuation must be performed from the region $\operatorname{Re} z > n/2$, that when n is even $\zeta(x, z)$ has simple poles at $z = 1, 2, \ldots, n/2$, and that when n is odd it has simple poles at $z = (n/2) - r$, $r = 1, 2, 3, \ldots$. One may furthermore conclude that $\zeta(x, z)$ is always well behaved at $z = 0$.

By writing $1/\Gamma(z)$ in the form $z/\Gamma(z+1)$, one also easily sees that

$$w(x) = -\frac{1}{2} i \operatorname{tr} \left[\frac{\partial}{\partial z} \zeta(x, z) \right]_{z=0} \tag{90}$$

$$W - S = -\frac{1}{2} i \zeta'(0) + \text{const.}, \tag{91}$$

$$\zeta(z) \equiv \int \zeta(x, z) \, d^n x. \tag{92}$$

When the zeta function is used, the 1-loop divergences are seen to be completely regulated away. Equation (84) may now be combined with Eq. (91) to yield

$$\langle T^{\mu\nu} \rangle - T^{\mu\nu}_{\text{classical}} = -i \frac{\delta \zeta'(0)}{\delta g_{\mu\nu}}. \tag{93}$$

In the special case in which $m = 0$ and $V = \frac{1}{4}[(n-2)/(n-1)] R$ [see Eq. (74)], the classical theory is conformally invariant and $g_{\mu\nu} T^{\mu\nu}_{\text{classical}} = 0$. One can then show [10] that

$$g_{\mu\nu}(x) \frac{\delta \zeta(z)}{\delta g_{\mu\nu}(x)} = z \operatorname{tr} \zeta(x, z), \tag{94}$$

$$g_{\mu\nu}(x) \frac{\delta \zeta'(z)}{\delta g_{\mu\nu}(x)} = \operatorname{tr} \zeta(x, z) + z \frac{\partial}{\partial z} \operatorname{tr} \zeta(x, z), \tag{95}$$

whence it follows that

$$\langle T_\mu{}^\mu \rangle = -i \operatorname{tr} \zeta(x, 0). \tag{96}$$

Inserting (21) into (89), carrying out the necessary analytic continuation, and performing a sufficient number of integrations by parts, one finds that $\zeta(x,0)$ is completely determined by the behavior of the integrand at $s=0$ and, in fact, that

$$\langle T_\mu{}^\mu \rangle = \begin{cases} 0 & \text{when } n \text{ is odd} \\ (4\pi)^{-n/2} g^{1/2}(x) \operatorname{tr} a_{n/2}(x,x) & \text{when } n \text{ is even} \end{cases}. \quad (97)$$

This is the well known *trace anomaly*. Other anomalies can be computed in analogous fashions.

The zeta-function method is not so useful for calculations involving two loops or higher. For these, however, one can fall back on dimensional regularization, in which n in Eq. (81) is regarded as a complex variable and the appropriate Feynman integrals are defined by analytic continuation from regions sufficiently far in the left-half n-plane. In the 1-loop case dimensional regularization gives physical results identical to those of zeta-function regularization.

Whether any of these lovely results will remain of interest to future generations remains to be seen. It is somewhat ironic to note that many of them were obtained as byproducts of the effort to develop a quantum theory of gravity. But it is in perturbation theory that propagators are mainly used, and quantum gravity — conventional quantum gravity, that is — is not a perturbatively renormalizable theory. Whether it can make sense as a nontrivial continuum limit of a lattice quantum theory is presently being studied with the aid of supercomputers. But I have no news to report from that effort at this time. Thank you for listening.

References

[1] J. L. Synge, *Proc. London Math. Soc.* **32**, 241 (1931).
[2] J. H. Van Vleck, *Proc. Nat. Acad. Sci., U. S. A.* **14**, 178 (1928). See also S. Lipschitz, *Bull. Sci. Math.* [1], **4**, 97 (1873).
[3] C. Morette, *Phys. Rev.* **81**, 848 (1951).

[4] B. S. DeWitt and R. W. Brehme, *Annals of Physics (N. Y.)* **9**, 220 (1960). See also J. M. Hobbs, *Annals of Physics (N. Y.)* **47**, 141 (1968).
[5] C. M. DeWitt and B. S. DeWitt, *Physics* **1**, 3 (1964).
[6] N. Bohr and L. Rosenfeld, *Kgl. Danske Videnskab. Selskab, Mat.-Jys. Med.* **12** (8) (1933).
[7] I learned this in a class on classical mechanics taught by Philipp Frank at Harvard in 1941, but I do not have a reference.
[8] B. S. DeWitt, in *Foundations of Quantum Mechanics* (1970 Varenna Lectures, Corso, IL), ed. B. d'Espagnat, Acadamic Press, New York (1971).
[9] R. E. Peierls, *Proc. Roy. Soc. (London)* **A214**, 143 (1952).
[10] B. S. DeWitt, in *Relativity, Groups and Topology II*, eds. DeWitt and Stora, North Holland, Amsterdam (1984).
[11] B. S. DeWitt, in *Gravitation: An Introduction to Current Research*, ed. L. Witten, Wiley, New York (1962).
[12] B. S. DeWitt, *Phys. Rev.* **160**, 1113 (1967); **162**, 1195, 1239 (1967).

Supported in part by a grant from the U. S. National Science Foundation.

TRIBUTE TO JULIAN SCHWINGER

Walter Kohn

University of California at Santa Barbara

It is a melancholy privilege for me to take part in this symposium in honor of my venerated teacher, Julian Schwinger.

All of us here know that his brilliant scientific insights and methodologies have left deep imprints across the entire spectrum of theoretical physics, both pure and applied. No doubt his most outstanding work was his monumental relativistically covariant renormalization theory of quantum electrodynamics; other areas which he substantially reshaped include quantum gauge theories, whose significance he was one of the first to realize; nuclear physics — beginning with his first papers written as a teenager and in which he offered perhaps the first comprehensive lecture course; the theory of waveguides, a powerful reformulation during World War II in terms of tensor Green's functions and variational principles; scattering theory; particle accelerators; and, very broadly, the structure of elementary particle theory.

As my own main interests, after about 6 years devoted to scattering theory, and some nuclear and particle physics, moved into the areas of solid state physics and density functional theory of matter, I do not feel adequately qualified to speak here in depth about Julian's researches. Instead I would like, in homage to his memory, to speak for a few minutes, from my perspective as an early graduate student and postdoc, about Julian as one of the greatest and most influential physics teachers of his generation. Following that I would like to tell you a little about DFT of electronic structure of matter, an area in which I have worked for about 30 years and which is slightly related to Julian's work in the early 80's on the statistical model of the atom.

Julian had altogether more than 70 Ph.D. students and I would guess of the order of 20 postdocs. At his 60th and 70th birthday celebrations, between his colleagues and former postdocs and students, I had a sense of experiencing in one place and at one time much of the history of post World War II theoretical physics.

Abe Klein has already described how he experienced Julian as a teacher. A few days ago, I talked with Quin Luttinger who was at M.I.T. at about the same time. Everyone experienced him differently, but everyone agrees that he was superb.

There were two distinct modes: Schwinger as a formal lecturer in the classroom and in seminars, and Julian as a mentor of budding theorists, typically just 5 or 6 years younger than he — in those early days.

Attending one of his formal lectures was comparable to hearing a new major concert by a very great composer, flawlessly performed by the composer himself. For example, his historic graduate courses on nuclear physics and on wave-guides given in the late 40's consisted largely of exciting original material. Furthermore both old and new material were treated from fresh points of view and organized in magnificent overall structures. The delivery was magisterial, even, carefully worded, irresistible like a mighty river. He commanded the attention of his audience entirely by the content and form of his material, and by his personal mastery of it, without a touch of dramatization. Interaction with the audience was as rare as in a formal concert. Crowds of students and more senior people from both Harvard and M.I.T. attended and, knowing his nocturnal working habits, I found the price of often having to wait 10, 20, 30 minutes for his arrival quite trivial in comparison with what he gave us. I felt privileged — and not a little daunted — to witness physics being made by one of its greatest masters. Each of these two courses had a tremendous influence on the shape of their respective fields for decades to come, as did other later Schwinger courses such as quantum mechanics and field theory.

I come now to Julian as mentor of more of the next generation of leading theoretical physicists than any of his contemporaries, I venture to say more than any two or three taken together.

It was easy to see why so many of us were drawn to him. Just a few years older than we, he appeared to be a blessed eighth wonder of the world, whose mind and famous files contained the answers to most questions we might think of, and who was among the 3 or 4 theorists most successfully pushing back the frontiers of the unknown and the paradoxical. In my particular case, my former teacher Leopold Infeld had specifically urged me to try to be accepted by him. Julian was full of good problems and the means with which to attack them. He accepted virtually everyone who wanted to work with him, during my day about 10 in all. A common, pre-existing love of variational principles, gave me a fortunate entrée to his circle and provided a durable scientific bond for the next years.

We all knew that he was one of the main contenders for the great prize of re-shaping quantum electrodynamics into a usable theory and were generally content to struggle with more mundane or peripheral problems. Arranging to meet with him was devilishly hard, but when it happened — a few times per year — I found him most generous with his time and brilliant in his judgments and suggestions. It was during these meetings, sometimes more than 2 hours long, that I learned the most from him. He had a large old-fashioned office in the old Jefferson building. In one corner, at a desk, sat Harold Levine calculating away on intricate classical wave problems, totally oblivious to what was going on around him. Drifting in and out were other students anxious to catch Julian. Frequently, Herman Feshbach came over from M.I.T. to talk about nuclear forces. A few times Freeman Dyson and Richard Feynman dropped in to talk about quantum electrodynamics. Once a letter or preprint from Tomanaga arrived and Julian said he was nervous to open it, so often had Tomonga's thinking been almost the same as his. What great fortune for us to be there at such a time!

And what did I carry away with me from my two years with Schwinger as his graduate student and two years as his postdoc? The self-admonition to try and measure up to his high standards; to dig for the essential; to pay attention to the experimental facts; to try to say something precise and operationally meaningful, even if — as is usual — one cannot calculate everything *a priori*; not to be satisfied until one has embedded his ideas in a coherent, logical, and esthetically satisfying structure.

Lastly some words of deep personal gratitude and of regret. I cannot even imagine my subsequent scientific life without Julian's example and teaching. I deeply appreciated it when he dropped me a line saying I was doing a good job at the Institute for Theoretical Physics in Santa Barbara. What caused me regret and pain over many years, and especially now, is that I had so little success in coming personally close to this shy, complex, and — I believe — lonely man, for whom I had and have the warmest feelings.

(Following these personal remarks Professor Kohn presented a review of Density Functional Theory in Physics and Chemistry (see *Density Functional Theory*, edited by E. K. V. Gross and R. M. Dreizler, Plenum Press, New York, 1995, pp. 3–10)).

OVERVIEW OF DENSITY FUNCTIONAL THEORY

W. Kohn

*Department of Physics, University of California,
Santa Barbara, CA 93106*

Introduction

I will first offer some "warm-up exercises" for density functional theory (DFT). After that I will present some very recent work which promises to allow Density Functional calculations for systems of very many (10^3–10^5) atoms.

I. Warm-up Exercises

Modern DFT for ground states is now almost 30 years old. The problem of dealing with a system of N interacting electrons in an external potential $v(r)$ is traditionally expressed by the $3N$-dimensional Schrödinger equation for the wave-function $\Psi(r_1, r_2, \cdots r_N)$. DFT recasts this problem in terms of the electronic density distribution, $n(r)$ and a universal functional of the density $E_{xc}[n(r)]$. (Formal DFT can be regarded as exactification both of Thomas-Fermi theory and of Hartree theory.) The problem of solving — necessarily approximately — the *many electron* Schrödinger equation is replaced by the problem of finding sufficiently accurate approximations to $E_{xc}[n(r)]$ and then solving appropriate *single electron* equations.

I believe that formal DFT would have been of very little interest if there had not been a simple and very practical approximation for E_{xc}, the local density approximation (LDA), which has yielded surprisingly accurate results:

$$E_{xc}^{\text{LDA}}[n(r)] \equiv \int n(r)\epsilon_{xc}[n(r)]dr, \qquad (1)$$

where $\epsilon_{xc}(n)$ is the very accurately known exchange-correlation energy per particle of a *uniform* electron gas of density n. With the single input of $\epsilon_{xc}(n)$, a function of one variable, the ground state energies and density distributions, $n(r)$, of any system can be easily calculated in the LDA with accuracies typically in the range of 10^{-3} to 10^{-1}, depending on the system and on the question asked.

The LDA is, by definition, exact for a homogeneous system, and arbitrarily accurate for a system of sufficiently slowly varying density. In real atomic, molecular condensed matter systems, $n(r)$ is a rather rapidly varying function of r. To allow for these gradients, $\nabla n(r)$, various corrections have been devised which reduce the errors of the LDA typically by factors of 3–5.

Many important generalizations of DFT and combinations with other theoretical concepts have been developed, some of which will be briefly mentioned below. It is my sense that at the present time DFT is the method of choice for systems consisting of many ($\gtrsim$ about 5) atoms and for smaller systems, when moderate accuracies are sufficient. DFT has for many years been used for calculating ground state properties of solids. In recent years, as a result of new gradient corrections of the LDA, DFT yields molecular binding energies of about the right accuracy (0.5–1.0 eV) needed for chemical applications, so that it is now receiving increasing attention by quantum chemists. Among other rapidly evolving areas I mention the combination of DFT with molecular dynamics for questions of structure and ion-dynamics, and the applications of thermal DFT to high temperature plasmas. As for other relatively recent developments such as time-dependent DFT, DFT of excited states, DFT of orbital magnetism, DFT of superconductors, etc., it is in my view too early to predict their eventual impact on physics and chemistry.

I shall now try to make brief reference to the key elements and applications of DFT. For original references and further details see Refs. 1–4.

The Hohenberg-Kohn (HK) Theorem

Consider a system of N interacting electrons in a non-degenerate ground state associated with an external potential $v(r)$.

Lemma: The ground state density, $n(r)$, uniquely determines the potential $v(r)$ (to within an additive constant).

Since with $v(r)$ the full Hamiltonian is known, this rigorous lemma (which can be easily purged of its restriction to non-degenerate ground states) has the consequence that $n(r)$ completely determines *all* properties of the system, such as all eigenfunctions $\Psi_j(r_1, \cdots r_N)$ and eigenvalues E_j, Green's functions such as $G(r, r', \omega)$, response functions, thermal properties, etc.

HK Variational Theorem

There exists a functional $F[n'(r)]$ defined for all non-degenerate ground state densities $n'(r)$ such that, for a given $v(r)$, the quantity

$$E_{v(r)}[n'(r)] = \int v(r) n'(r) dr + F[n'] \qquad (2)$$

has its unique minimum for correct ground-state density, $n'(r) = n(r)$, associated with $v(r)$. The physical meaning of $F[n']$ is

$$F[n'(r)] \equiv (\Psi_{n'(r)}, (T + U)\Psi_{n'(r)}) \qquad (3)$$

where $\Psi_{n'(r)}$ is the ground state associated with $n'(r)$, and T and U are the kinetic energy and Coulomb repulsion operators.

M. Levy and E. Lieb have independently shown by specific example that not all well-behaved functions $n(r)$ can be realized as ground state densities. However, this so-called v-representability problem has, so far, not caused any practical difficulty.

The same authors have also pointed out that $F[n'(r)]$ can be defined as

$$F[n'(r)] \equiv \min_{\Phi_{n'(r)}} (\Phi_{n'(r)}, (T+U)\Phi_{n'(r)}), \tag{4}$$

where the minimum is taken over *all* normalized antisymmetric wave function $\Phi_{n'}$ (not necessarily solutions of a Schrödinger equation) which give rise to a density $n'(r)$. This "constrained search" definition of $F[n'(r)]$ is valid for a larger class of densities, $n'(r)$, than those which are physical, non-degenerate ground states.

Kohn-Sham (KS) Equations

The Hohenberg-Kohn variational principle can be recast in the form of exact single particle self-consistent equations, similar to the Hartree equations:

$$\left(-\frac{\hbar^2}{2m}\nabla^2 + v_{\text{eff}}(r)\right)\psi_j(r) = \epsilon_j \psi_j(r)$$

$$n(r) = \sum_{j=1}^{N} |\psi_j(r)|^2 \tag{5}$$

$$v_{\text{eff}}(r) = v(r) + \int \frac{n(r')}{|r-r'|}dr' + v_{xc}(r).$$

Here $v_{xc}(r)$ is the local exchange-correlation potential, defined as

$$v_{xc}(r) = \frac{\delta E_{xc}[n(r)]}{\delta n(r)}$$

$$E_{xc}[n(r)] = F[n(r)] - \frac{1}{2}\int \frac{n(r)n(r')}{|r-r'|}dr' - T_s[n(r)], \tag{6}$$

where $T_s[n(r)]$ is the kinetic energy of *non-interacting* electrons with ground state density $n(r)$.

In the LDA, v_{xc} becomes simply

$$v_{xc}^{LDA}(r) = \frac{d}{dn}(\epsilon_{xc}(n)n)\bigg|_{n=n(r)}. \tag{7}$$

Solution of the Eq. (5) yields $n(r)$ and allows calculation of the ground state energy

$$E = \sum_{j=1}^{N} \epsilon_j - \frac{1}{2}\int \frac{n(r)n(r')}{|r-r'|}dr' - \int v_{xc}(r)n(r)dr + E_{xc}[n(r)]. \tag{8}$$

Generalizations

The HKS theory of the previous section has been generalized to the following systems.

a. *Multicomponent Systems.* E.g. electrons and holes in semiconductors. The correspondence $n(r) \to v(r)$ becomes $\{n_1(r), n_2(r)\} \to \{v_1(r), v_2(r)\}$.

b. *Spin Paramagnetism.* E.g. atoms with odd values of Z. The appropriate densities are $n(r)$ and $m(r) \equiv n_+(r) - n_-(r)$, where n_+ and n_- are spin-up and spin-down densities. The corresponding "potentials" are $v(r)$ and the magnetic field $B_z(r)$.

c. *Orbital Magnetism.* The pertinent "densities" are $n(r)$ and $\vec{j}_p(r)$ the paramagnetic (gauge-dependent!) current density. The corresponding "potentials" are $v(r)$ and $\vec{A}(r)$.

d. *Excited States.* One calculates $\bar{E}$, the *average* of the M lowest eigenvalues. The appropriate density is $\bar{n}(r) = (n_1(r) + \cdots n_M(r))/M$ and the corresponding potential is $v(r)$.

e. *Superconductivity.* The appropriate densities are the conventional density, $n(r) = \sum_\sigma \langle \psi_\sigma^*(r)\psi_\sigma(r)\rangle$; and the pair-density $\tilde{n}(r) = \sum_\sigma \langle \psi_\sigma(r)\psi_{-\sigma}(r)\rangle$; the corresponding potentials are $v(r)$ and a (real or fictitious) pair-potential, $D(r)$.

f. Thermal Ensembles. For a finite temperature grand canonical ensemble the ground-state energy is replaced by the grand potential Ω, and the functional to be minimized has the form

$$\Omega^\beta_{v(r)-\mu}[n'(r)] \equiv \int (v(r) - \mu)n'(r)dr + F^\beta[n'(r)], \qquad (9)$$

where $\beta = (kT)^{-1}$ and μ is the chemical potential. The functional F now depends parametrically on the temperature. The relevant density is $n'(r)$ and the corresponding potential is $(v(r) - \mu)$.

Thermal ensembles are free of certain technical problems which exist for ground states: the issue of degeneracy is irrelevant; the arbitrary additive constant in $v(r)$ does not enter $v(r) - \mu$; and finally v-representability of given densities $n(r)$ is much broader at finite temperatures.

From the minimal property of Ω, Eq. (9), a set of self-consistent equations analogous to the zero-temperature KS equations, (5), can be derived, involving a temperature dependent exchange correlation functional analogous to the $E_{xc}[n'(r)]$ of ground state theory.

g. Time Dependent DFT. A HK-type theorem and KS-like equations have been derived for systems with a time dependent external potential $v(r,t)$ and corresponding density $n(r,t)$, $t > 0$. The general case is formidable because the unknown functional depends parametrically on the initial state. However a practical LDA has been developed for linear response, i.e. when $v(r,t) = v_0(r) + \lambda v_1(r,t)$ and λ is small.

Approximations

Unquestionably the basic approximation for E_{xc} is the local density approximation (LDA), Eq. (1), which requires only a knowledge of the exchange-correlation energy $\epsilon_{xc}(n)$ of the uniform electron gas. (Analogous LDA's exist for the generalizations mentioned in the last section.)

Improvements of the LDA can be grouped in several categories:

(a) Formal gradient expansions, stopping usually at $(\nabla n(r))^2$, i.e.

$$E_{xc}[n(r)] = \int \epsilon_{xc}(n(r))n(r)dr + \int g_2(n(r))|\nabla n(r)|^2 dr + \cdots . \quad (10)$$

In applications to atoms, molecules and condensed matter these formal expansions have generally been disappointing, suggesting an asymptotically convergent series with the best results obtained by keeping just one term (LDA) or two terms. Addition of the gradient term generally does not significantly improve the accuracy.

(b) Semiphenomenological gradient functionals.

The LDA approximation of E_{xc} can be written in the form

$$E_{xc}^{LDA} = \int f_1(n(r))dr, \quad (11)$$

where $f_1(n)$ is a function of the single variable n.

This may be generalized to

$$E_{xc}^{(1)} = \int f_2(n(r), |\nabla n(r)|)dr, \quad (12)$$

where f_2 is a function of 2 variables, $n(r), |\nabla n(r)|$. The formal gradient expansion is of such a form. More useful functional forms for $f_2(x, y)$ have been obtained by a combination of physical principles and phenomenological fitting to exact experimental or theoretical results.

(c) Approximations to the exchange-correlation hole.

In an electron gas — with or without Coulomb repulsions — each electron, located at say r_1, is surrounded by a deficiency in electron density at neighboring points r_2; $n_{xc}(r_1, r_2)$, which integrates to -1. The exchange correlation energy can be expressed exactly as

$$E_{xc}[n] = \frac{1}{2} \int dr_1 dr_2 \frac{1}{|r_1 - r_2|} n(r_1) \bar{n}_{xc}(r_1, r_2), \quad (13)$$

where $\bar{n}_{xc}$ is the exchange-correlation hole suitably averaged over the interaction coupling constant.

The hole function $\bar{n}_{xc}$ is accurately known for a homogeneous gas, integrates to -1 in all systems and has a known functional form for $r_2 \to r_1$ (Coulomb cusp). Various approximations have been explored, with mixed results, to describe this function.

Density Functional Theory and Molecular Dynamics

In the previous sections we have implicitly assumed that we are dealing with problems of *given* $v(r)$, i.e. known positions of the atomic nuclei. However for many physical systems such as multi-atomic molecules, liquids, solutions etc. the atomic positions are not known *a priori* but determined self-consistently by the electronic density $n(r)$.

An efficient way of carrying out this program was first proposed by Car and Parrinello who simulate a fictitious annealing process which, starting at high temperatures allows the system to cool into a zero temperature equilibrium configuration. This method combined with the LDA has made possible calculations of previously inaccessible equilibrium configurations of systems with up to $\sim 10^2$ atoms.

Applications and Work-To-Do

In this brief section I will not attempt the impossible, i.e. to offer a critical review of the multitude of applications in Physics and Chemistry. For this I refer again to recent books and review [1, 2, 3, 4] and to several of the subsequent papers of this conference.*
Nor will I attempt to make a complete list of remaining problems. Instead I will only present some personal perceptions.

(a) DFT has found many useful applications when moderate accuracies (typically in the range 10^{-3} to 10^{-1}) are required. It is *not*

---

* "This conference" here refers to NATO ASI series B337 (Ciocco, 1993) — editor's note.

a precision method which, in principle, can be pushed to arbitrary accuracy.

(b) Among its most useful applications have been the electronic structures of periodic solids for which — in most cases — there exist no competitive alternatives. Solutions of the KS equations yield not only total energies and density distributions but also KS energy bands. Although the latter have no strict physical significance, they very frequently have provided important insights into electronic structures.

(c) While I see no reason why DFT should not *in principle* describe complex solids like fluctuating valence compounds, Mott insulators etc., it is not clear to me to what extent the simple descriptions of many body effects, such as the LDA and its refinements (see above) can contribute to their understanding.

(d) Many important applications, like surface properties and molecular binding energies, involve electrons in a negative energy (or "forbidden") regime, i.e. referring to the KS equations (5) where $(\epsilon_j - v_{\text{eff}}(r)) < 0$. This regime is qualitatively different from that of a uniform gas. I believe that the question of the nature of many-body effects in this regime is a major challenge and that even moderate progress could have great impact on chemical applications.

(e) Some many-body effects are *truly* non-local, e.g. dispersion forces behaving as r^{-n} for larger r. Here also further progress will surely not come from refinements of the LDA.

(f) The related problems of excited states and time-dependent linear response have recently seen some progress. I hope this is the beginning of much broader future developments, but I am not sure. Perhaps the "good luck" which we have had with the LDA in ground-states will not hold up for these problems.

(g) In recent years a number of quantum chemists have recognized that DFT offers the possibilities to study *larger* molecules or clusters, with a number of atoms, N, up to ~ 50–100, and with accuracies close to what is needed for understanding chemical reactions. Traditional methods of quantum chemistry typically do not go beyond $N \sim 5\text{–}10$. Improving DFT accuracy by another factor of 2 would make a major difference.

(h) Finally, even the now traditional Car-Parrinello scheme hits a practical barrier at some number N of atoms, which currently is about 50–100, because computing time still rises as N^3 or N^4. Very recently progress has been made (see Part II) in developing approaches behaving linearly in N, which show promise of handling truly large systems with $N \sim 10^3\text{–}10^5$.

II. Density Functional/Wannier Function Theory for Systems of Very Many Atoms

A very basic property of many-electron systems in equilibrium is that properties near a point r_0, such as $n(r_0)$, depend significantly on the form of the potential $v(r)$ *only* for points r near r_0, as well as on the overall chemical potential, μ. I have elsewhere called this property the *locality property* of quantum mechanics. While it is easy to verify this property in representative examples and I am persuaded of its wide applicability, I have no idea how to prove it mathematically from the $3N$-dimensional partial differential Schrödinger equation.

Related to this conjecture are the locality properties of so-called generalized Wannier functions [5], defined as follows. Consider the single particle Schrödinger equation corresponding to a potential $v(r)$ plus boundary conditions. Assume further that there is a "band" of adjacent N eigenvalues, ϵ_j, which are well separated from all other eigenvalues. Then we claim that the corresponding N eigenfunctions, φ_j, which may or may not be localized, can be replaced by means of a unitary transformation by N orthonormal generalized Wannier function, w_ℓ, which are well localized near sites R_ℓ and whose shape depends only on the form of the potential $v(r)$ for r near R_ℓ. If in a many-electron state all N states φ_j are occupied (we neglect for the moment all other states), the corresponding Slater determinant of the φ_j can be replaced by the determinant of the GWF's w_ℓ.

Now the form of a particular extended $\varphi_j(r)$ near a point r_0 can depend strongly on the potential $v(r)$ for r far from r_0. I call this property of $\varphi_j(r)$ *far-sightedness*. By contrast, the form of each $w_\ell(r)$ depends only on the form of $v(r)$ for r near R_ℓ. I therefore call the $w_\ell(r)$ *near-sighted*. Any property of the many-electron system near a point r_0 can be expressed in terms of the well-localized w_ℓ with R_ℓ near r_0; for example:

$$n(r_0) = \sum_\ell w_\ell^2(r_0), \qquad (14)$$

where effectively the sum goes only over those values of ℓ for which R_ℓ is near r_0.

We can take advantage of the above short-sightedness as follows. In the self-consistent KS-scheme (see Eq. (5) ff.) we start from an assumed $v_{\text{eff}}(r)$ and need to find $n(r)$ as well as $\sum_1^N \epsilon_j$ for calculation of the total energy. Instead of calculating the far-sighted φ_j, a task which grows as N^3 or N^4, we calculate the near-sighted, well-localized w_ℓ which depend only on their neighborhood and whose calculation therefore requires only of order N^1 steps. Details are given in [5].

Approaches similar in spirit but different in specifics can be found in [6–8]. These papers strongly suggest that systems of 10^3 to 10^5 atoms will soon be susceptible to DF calculations.

Let me end by two remarks: One is that the "near-sightedness" of GWF's corresponds closely to the chemist's concept of transferability of bonds from one compound to another. Secondly, exponentially localized GWF's do not exist if there is no energy gap, e.g. for metals. Nevertheless, the locality property of quantum mechanics holds even for gapless systems, though in weaker, algebraic ($r^{-\gamma}$) form. At this time I do not know what is the most fitting way to treat such gapless systems in a N^1 scheme.

Acknowledgments

I am very pleased to acknowledge very stimulating discussions with M. Teter concerning Part II, and the support of NSF Grant DMR93-0801.

References

[1] R. G. Parr and W. Yang, *Density Functional Theory of Atoms and Molecules* (Oxford, 1989).
[2] R. O. Jones and O. Gunnarsson, "Density Functional Theory, Its Applications and Prospects," *Rev. Mod. Physics* **61**, 689 (1989).
[3] S. B. Trickey, ed., "Density Functional Theory of Many Fermion Systems," *Adv. in Chem.*, **21** (Academic Press, 1990).
[4] R. M. Dreizler and E. K. U. Gross, *Density Functional Theory* (Springer, 1990).
[5] W. Kohn, *Chem. Phys. Lett.* **208**, 167 (1993).
[6] D. C. Allan and M. P. Teter, *J. Am. Ceram. Soc.* **73**, 3247 (1990).
[7] W. Yang, *Phys. Rev. Lett.* **66**, 1438 (1992).
[8] G. Galli and M. Parrinello, *Phys. Rev. Lett.* **69**, 3547 (1992).

Schwinger with Robert Oppenheimer at Eagle's Nest (Oppenheimer's residence at Berkeley) in 1948.

Schwinger (center) at Lake Como, Italy, in 1949, with (left to right) I. I. Rabi, S. White, E. McMillan, and G. Placzek.

Schwinger at Schladming, Austria, in 1975 with (left to right) H. Mitter, Clarice Schwinger, M. Mitter, J. Klauder, and H. Latal. Mrs. Schwinger once told the editor, "Life with Julian was fun. He cared about and was interested in so many things."

JULIAN SCHWINGER MEMORIAL TRIBUTE

David Saxon

University of California at Los Angeles

Introductory Remarks: The Early Years

My name is David Saxon. On behalf of the UCLA Physics and Astronomy Department, I welcome you to this ceremony in honor of Julian Schwinger.

Julian died just over three months ago, on Saturday July 16, at the age of 76. The cause of his death was pancreatic cancer. His decline was rapid, but of course he continued working to the very end, seeking perfection as always. Of course.

Despite his many honors (among these the Nobel, Einstein and Nature of Light awards) Julian remained what he had always been, an unassuming, shy, gentle and intensely private person; well-read and well-traveled but surprisingly unworldly; modest about everything except his physics (as Clarice used to say on occasion); generous with his time and his help (once you found him); a lover of music, good food and good wine (he was part-owner of an excellent small winery); surprisingly athletic, an avid skier and tennis player, singles (he beat me handily). He had the rare gift of silence. His humor was too subtle for small talk but he was an attentive and deeply thoughtful listener (some times disturbingly thoughtful).

I speak from long experience, very long, for Julian and I met 51 years ago, in 1943, at the wartime MIT Radiation Laboratory. I was an MIT graduate student working on my dissertation under Slater, he was at 25 already a legendary figure, an acknowledged prodigy, a genius. As Sam Goudsmit put it to me, "Julian would have already won the Nobel Prize if he had been born at a more propitious time," by which Sam meant in time to have participated in the explosively exciting development of quantum mechanics in the 1920's.

Although there had been no science at all in his background, Julian became committed to physics at an unusually early age. He once remarked, by way of explanation, that he had been reading the family *Encyclopedia Britannica* straight through; when he came to the letter P and physics. That did it. Largely self-educated, he published his first paper, with Halpern, in 1935 at the age of 17, while an undergraduate at CCNY. His abilities caught the attention of Rabi, who brought him to Columbia where he published his second paper, with Motz, later that same year. Both papers were brief and rather minor. His next paper, on the scattering of neutrons by magnetic materials, and his first solo publication, was another matter entirely. Submitted to Physical Review when he was still 18, it was an important piece of work, mature and elegant. The Schwinger techniques, his characteristic mastery of the subject and his physical insight were on full display for the first time.

His subsequent development was rapid. "He emerged", Fred Seitz writes, "like a comet as a brilliant self-guided genius." At Columbia, Wisconsin and Berkeley, alone or in collaboration with (in alphabetical order) Corbin, Gerjuoy, M. Hamermesh, Oppenheimer, Rabi, Rarita, Sachs and Teller among others, he worked on $n-p$ scattering, the quadruple moment of the deuteron, tensor forces, light nuclei, pair emission, meson theory of nuclear forces, the spin-dependence of nuclear forces and more.

When war came (The War) he was urged to join the Manhattan Project, but he chose not to work on the development of nuclear weapons despite his nuclear expertise. He came instead to the MIT radiation Laboratory to work on radar, a subject about which he knew little. His war-time work was important in two respects, at least. First, in respect to the war effort, Julian developed a powerful and systematic approach to the analysis and solution of the very practical problems of the propagation of microwaves in wave guide structures and of the properties of wave guide junctions. His work had applicability to the construction of working radar systems such as that used in the detection of German submarines in the Bay of Biscay when they surfaced at night to recharge their batteries. Second, in respect to science, the techniques he developed played an essential part in his subsequent work, specifically — as Julian pointed out 50 years later in one of his last papers — his mastery of Green's functions began with his work on microwave propagation.

After the war, Julian joined the faculty at Harvard, where he became a full professor at 29. Between 1948 and 1950 came the contributions which led to his Nobel Prize, a prize he shared with Feynman and Tomonaga. During that period he also assumed his extraordinarily influential role as teacher and mentor to an eventual 70 or more Ph.D. students. I am pleased that three of his students from his Harvard days are here to join us in honoring him. They are Paul Martin, Kenneth Johnson and Stanley Deser. They will speak in that order and then Robert Finkelstein will speak on Julian's days here at UCLA.

JULIAN SCHWINGER — PERSONAL RECOLLECTIONS

Paul C. Martin

Harvard University

We're gathered here today to salute Julian Schwinger, a towering figure of the golden age of physics — and a kind and gentle human being. Even at our best universities, people with Julian's talent and his passion for discovery and perfection are rare — so rare that neither they nor the rest of us know how to take best advantage of their genius. The failure to find a happier solution to this dilemma in recent years has concerned many of us. It should not becloud the fact that over their lifetimes, few physicists, if any, have surmounted this impedance mismatch more effectively than Julian, conveying not only knowledge but lofty values and aspirations directly and indirectly to thousands of physicists.

Since his death three months ago, Julian's many contributions to nuclear and atomic physics and to quantum electrodynamics have been recounted, as have the scores of techniques and insights he originated that are now in the tool-bag of every theoretical physicist. Later this afternoon, some of his contributions to gauge theories, general relativity, and particle physics will be described.

The scope and significance of his discoveries are monumental, and the breadth and beauty of his lectures and writings are legendary.

Reminiscences of his creativity as a researcher and his brilliance as a lecturer liken him to a great musical composer and performer. The seventy or so doctoral students he supervised (and their hundreds of descendants) are only some of those he influenced. When Julian offered a course, a large segment of the community attended and took notes.

Significant portions of many of today's texts on nuclear physics; atomic physics; optics, electromagnetic theory, and wave guides; statistical physics; quantum mechanics; and quantum field theory have been strongly influenced by techniques, approaches, and examples Julian introduced in his lectures. Many "standard" treatments can be traced back to lecture notes assembled and embellished by Julian's students and widely disseminated throughout the community.

Among many examples are his treatments of: (i) effective range theory; (ii) scattering; (iii) tensor forces and quadrupole moments; (iv) variational principles; (v) Green's functions for classical and quantum fields; (vi) angular momentum in terms of oscillators; (vii) Coulomb states in momentum space and oscillators in external fields; (viii) commutators of currents, including the momentum current; and (ix) magnetic monopoles, higher spin particles, and gravitons. Others will add to this list.

During these past three months, stories of how Julian developed a profound understanding of physics and mathematics by studying the *ENCYCLOPEDIA BRITANNICA* also have been recounted, along with reminiscences about his less than punctilious behavior as an undergraduate and his rapid rise to stardom at Columbia, under Rabi's wing. By the age of twenty he was an impresario in a field for which the twentieth century has been the golden era. I'm delighted that some of his early coworkers are here and look forward to hearing more first hand stories from friends like Ed Gerjuoy, Morton Hamermesh, Harold Levine, and Nathan Marcuvitz who knew him at an earlier stage, while his fame was spreading.

The tales of his youthful genius were well known to those of us who had the good fortune to be his students in the late 1940s and 1950s, attending his original and inspirational lectures several times a week. Seated in the front rows of large, attentive audiences of students and faculty from Harvard and MIT, we watched and listened with a mixture of relish and reverence. The ritualistic features of these lectures have been noted; they began late, and ended later. Speaking eloquently, without notes, and writing with both hands, he expressed what was already known in new, unified ways, incorporating original examples and results almost every day. Interrupting the flow with questions was like interrupting a theatrical performance. The lectures continued through Harvard's reading period and then the examination period. In one course we attended, he presented the last lecture — a novel calculation of the Lamb Shift — during Commencement Week. The audience continued coming and he continued lecturing.

Seeking Julian's advice on thesis research also had idiosyncratic features. He usually reserved one afternoon a week for such discussions, and neither the starting time nor the duration of the discussion was fixed. Enthralled by Julian's way of expressing results, most of us tried to express our conclusions and problems in similar terms.

As members of college fraternities and sororities sometimes adopt a jargon and rituals, we graduate students took a certain sophomoric pride in our own patois. It gave our group a special cachet yet, in contrast with secret handshakes and patter, our rituals and jargon gave us an extra set of powerful tools and insights — Green's functions, functional derivatives, and variational principles — that improved and extended our understanding. Truth was expressed most concisely on eight pages in the 1951 *Proceedings of the National Academy of Sciences*.

When we get together, as we did to celebrate Julian's sixtieth birthday, we like to recall and romanticize these rituals. My own remarks at Julian's 60th birthday celebration, which are reproduced in Schweber's book on the history of quantum electrodynamics, have this tenor.

Unfortunately, these reminiscences have been seriously misconstrued. Some people have inferred that we were unaware of or unable or unwilling to use other languages. Some have even suggested that Julian enforced conformity and a boot-camp approach. As Abe Klein pointed out in remarks delivered at a symposium last month, these inferences are misguided. True, students often had to wait in line to see Julian, but there were also days when no one was waiting. We sought him out for advice infrequently, because we did not wish to impose on him and because we felt we should be able to make progress by ourselves and by consulting one another.

As to conversations we held with him as graduate students, he might frown when one of us drew a Feynman diagram, but we knew all about those diagrams, including how to generate them quickly and concisely from functional equations that bypassed Wick theorems and the like. Sensitive souls, recognizing that a frown was the most overt sign of displeasure Julian would ever display, might have refrained. But I and many others were not sensitive — and none of us were treated less warmly or generously as a result of such transgressions. Given our numbers, Julian's and Clarice's sincere concern for us, our spouses, and our progeny has been extraordinary.

Still, I and many others feared imposing upon him. Rather than present him with an invitation he might feel obliged to accept, we did not ask him to speak at my own sixtieth birthday celebration. To my surprise, delight, and embarrassment, Julian and Clarice did attend as members of the throng. I note that Walter Kohn expresses similar delight upon receiving a congratulatory note from Julian on Walter's contributions to the Institute for Theoretical Physics at Santa Barbara.

Julian and I and Clarice and my wife Ann spent a great deal of time together in Madison, Wisconsin during the summer of 1958.

Earlier in 1958, upon my return from Copenhagen and Birmingham, where I had carried out various calculations on interacting fermions and bosons, I turned to Julian with a number of questions and suggestions. He arranged with Bob Sachs for us to join Clarice and him at the University of Wisconsin where he was also lecturing and working on problems in particle physics. It was an eventful summer. I would carry out a field-theoretic calculations using our temperature-dependent Green's function approach, and he would generalize them, or he would generalize and carry out his own calculations to see how they went. We would talk about problems together, in his backyard or ours or on an excursion to Taliesin East or on the lake, and then go off to work and write separately. Neither of us would have guessed that this paper would capture the attention of mathematicians who would speak of KMS analyticity.

Many visitors came through Madison that summer, and Clarice and Julian entertained most of them. Our own home was a dormitory for graduate students visiting Julian. One, who defended his Ph.D. thesis there, was Shelly Glashow. His committee consisted of Bob Sachs, Frank Yang, Julian, and me. A high point of the examination, oft recalled by Shelly and me, was a debate between Frank Yang and Julian. Frank was most unsympathetic to Julian's theory, which Shelly was investigating. Why, Frank asked, should anyone want a theory with more than one two-component neutrino — one in which the muon and electron had different lepton numbers and were associated with different neutrinos? Shelly and I expect that Frank may have forgotten that discussion.

Although statistical and condensed matter physics are not fields with which many people associate Julian's name, they should be. In 1961, Julian developed systematic techniques for treating quantum systems away from equilibrium — techniques that require additional Green's functions because there is no fluctuation dissipation theorem. This theory, also developed by Keldysh, is now being used to describe and analyze the behavior of microelectronic devices. More recently, he and Berthold Englert have studied atomic physics using the Fermi-Thomas approximation and the Casimir effect at finite temperatures.

In addition to seeing each other in Cambridge and Belmont, where everyone was always very busy, we spent many happy hours with the Schwingers in Paris in the spring of 1963. On one of these occasions, our eldest son, then two, had his first "happy hour" drinking an unattended alcoholic beverage that looked like water.

For many years, theorists from Harvard and MIT (Feshbach, Weisskopf, Low, Johnson, Glauber, Glashow, Coleman, and Jackiw) joined Julian for lunch twice a week at a local restaurant where the menu and Julian's selection were invariant but the discussions of physics wide-ranging.

Despite his fame, achievements, and ability, Julian was never content to rest on his laurels. He never stopped trying to make new discoveries, and he never stopped fretting about how to convey knowledge to others at all levels, from original research to basic graduate courses to pre-college education. Whatever his audience, he thought long and hard about how to make science comprehensible and attractive. He made a habit of staying home to prepare on evenings before he lectured. Walter Kohn put it this way: "What did I carry away with me from my years with Schwinger? The self-admonition to try and measure up to his high standards; to dig for the essential; to pay attention to the experimental facts; to try to say something precise and operationally meaningful even if — as is usual — one cannot calculate everything; not to be satisfied until one has embedded his ideas in a coherent, logical, and aesthetically satisfying structure."

Not through statements but through action Julian instilled in each of us a commitment to approach every subject we chose to study in a broad context, assuming as little as possible, and seeking new verifiable results; to place the highest value on discovery, presentation, and conveyance of knowledge; and not to waste time and energy on political maneuvering and political battling; to strive always to understand and extend, to unify and generalize; and to observe from new directions that reveal new insights.

How many of Julian's students recall, and hope to emulate these stern values, recognizing that the continual pursuit of difficult challenges is hardly a formula for happiness!

Curiously, I suspect that many of us have been able to follow the example he set with fewer disappointments. Not having his extraordinary talent and correspondingly demanding self-expectations, we have coped more easily with not succeeding in every ambitious undertaking or when others solved a problem we were hoping to solve.

The world will long remember Julian's great discoveries, and even his non-enduring theories are recorded for posterity. What we say here today will, and how others have described him personally may, soon be forgotten. But inferences that Julian's influence on physics was limited solely to his publications and discoveries — or even solely to his research and formal lectures — miss the point. We who had the opportunity to benefit from him personally know better. That is what has drawn us here today. We hope that he recognized, and want to be sure that Clarice recognizes the gratitude and affection so many of us feel but found it so awkward to express.

JULIAN SCHWINGER — PERSONAL RECOLLECTIONS

Kenneth Johnson

Massachusetts Institute of Technology

In this talk, I will describe the Julian Schwinger whom I remember as a teacher and friend during the time when I had the closest association with him and which was the formative period in my life as a theorist. This was the period from late 1952 to mid 1957. In order that I can do this, occasionally I also will have to tell you a little about me. At the end of the spring in 1957 I left Harvard and went to Los Alamos for a short time and then I went to the Bohr Institute. When I returned to Cambridge it was to MIT, which although close to Harvard still I didn't meet with Julian very often since it was a time when Julian had a large number of very talented students who wished to have some of his time. Thus it was then that our friendship continued but my close association with him was over. I would not attempt to describe here his research except briefly, even during this short period, so this will be a more personal account of my contact with him. I will divide my remembrances of Julian into my views of him as a superb teacher, as a fantastically creative mathematical physicist, as a preeminent theorist, and as a warm and good friend.

In September 1952, I arrived in Cambridge, Massachusetts at the Physics department of Harvard University as a beginning graduate student. I had obtained a reasonable background in classical physics and analysis as a student at a small engineering school (I.I.T.) in Chicago. I was fortunate to have had a very good Physics teacher there who subsequently become a member of your faculty, namely Burton D. Fried. For some reason the mathematics faculty at I.I.T. was at that time quite distinguished so I also had gotten a rather good foundation in analysis from several teachers. When I expressed an interest in graduate school, Fried suggested that I apply to Harvard because he said that Julian Schwinger was the outstanding teacher of the kind of theoretical physics that I expressed an interest in. I did as suggested and fortunately was admitted. Although at Harvard it was recommended by my faculty and graduate student advisers that I not begin my study of quantum mechanics with Schwinger's course, since I was nearly completely ignorant in that area, I not unreasonably disregarded the advice since I had arrived with the goal of learning it from him.

I soon discovered Schwinger's style of lecturing was unique. Without holding anything written in his hands, each step followed logically from the previous analogously to the way the notes in a sonata by Mozart follow uniquely one after another. Similarly, just as you would not interrupt a great pianist in the middle and ask him to repeat something, so one did not interrupt Julian with questions. At first, I was rather bewildered because although the Schwinger lectures were very clear and elegant, the information provided seemed to have very little relationship to what I found in most of the text books on quantum mechanics in the library (although at that time there were not many). The closest to the material discussed in the lectures seemed to be present in the book by Dirac, but even there the correspondence was mostly in some of the mathematical aspects. It was only near the end of the first semester when the correspondences finally began to become clear when the "action principle" was revealed. Indeed, then not only did the mysterious quantum theory come into full view, simultaneously its relationship to the canonical transformation theory of classical mechanics was expressed in a way so that most of what I had studied about the classical theory was also put in a form which made it appear much more natural and clear. Much of this material is available in the little book, *Quantum Kinematics and Dynamics* published by W.A. Benjamin, Inc. I do not know whether or not it is still in print. I believe that if it still is, it would be found to be an extremely valuable resource for students of quantum mechanics. I have loaned my copy many times and it is very well worn. During the remainder of that first year, we learned how to do most of the standard quantum problems using this formulation of quantum mechanics. I still have my notes of those quantum courses and they provide me with teaching materials which 40 years later still provide "novel" approaches to some of the classic examples of quantum systems. It was at this time that I first learned that Julian Schwinger was a true master of "variational principles" and at the same time I found out that these methods provided elegant mathematical resources for solving physics problems. In the third semester of this course we learned about "Green's functions" and how powerful these tools became in the hands of Julian Schwinger. I still believe that these methods are the most transparent ways of solving problems in many areas of theoretical physics. They appear in many places and Schwinger's legacy in this area is truly profound.

After these three semesters of quantum mechanics, I expected that we would move on to quantum field theory but unfortunately (for me) Schwinger was not there but was absent for a term on a sabbatical leave which he might have spent at home working, at least it was during this period that the famous series of papers entitled "Quantum Theory of Fields I-VI" began to appear. I spent this time learning how to calculate and repeated some of the calculations that many others had made during the late forties and early fifties. In the fall of 1954, Julian was in residence again and the next phase in my relationship began with him as my "thesis adviser." It was then that Julian became as aware of me as I was of him. At that time, as I recall, there was a mysterious dip in the number of students who wished to "work for Schwinger" so I did not experience the later legendary waits on Wednesday afternoon when he would patiently listen to and advise his students. I always found him to be very tolerant and helpful in giving his counsel which most often helped me to move onto the next stage of research. Thus I remember him both as a superb classroom teacher and as a very kind and helpful research adviser. It was at this time that Julian gave a series of lectures on further developments of his functional formulation of equations for Green's functions of quantum fields which was first presented in his famous National Academy papers of 1951. There the use of "sources" as functional variables was introduced, ordinary classical sources for bosonic fields and Grassman sources for fermionic fields. It was the functional differential equation version that is in integral form presently called functional integration. Using this many of the symmetry properties of the Green's functions can most transparently be gotten. This work alone would have been sufficient to make one famous as a mathematical physicist. I was later impressed to see how much of this material was rediscovered by others. Part of the thesis problem he gave me was to work out this formulation for scalar charged fields.

During this period at Harvard, Julian had an "assistant" whose duties I found out, at least when I later took the position, were to have lunch with Julian on Wednesday after he was finished with his students and carry out your own research. When in the spring of 1955 Paul Martin decided that the attractions of Tivoli were greater than those of Cambridge Common and decided to vacate the "assistant" position, I was lucky enough to be asked by Julian to take it up. To my surprise, I found out also that he thought I had done enough for my degree. Thus in the fall of 1955, I became Julian's "assistant." For the next two years, I had the wonderful experience of learning from Julian what he was thinking about on those Wednesdays when no one else would join the lunch. When others were around Julian usually listened unless as was often the case someone would be asking him for help with his current research. He seemed to enjoy being asked specific questions. On the other hand, he appeared to dislike being asked about his "opinions" of this or that. He didn't often say very much unless he was currently getting interested in a topic and wanted to get specific information about it. I remember in particular a couple of times when Julian discussed with me inconsistencies which could be produced through formal calculations with composite operators, what later came to be called "Schwinger terms." During this time it was suggested to me that I might investigate the gauge properties of Greens functions, using the functional formalism. One of Julian's most beautiful and still studied papers is the 1951 paper on vacuum polarization and gauge invariance. In it he introduced an elegant way to avoid problems of gauge invariance for an electron field coupled to an external electromagnetic field. He used a proper time first quantized formalism for the charged particle as a tool for constructing the field theoretic Green's functions. He wanted me to investigate the gauge questions when the electromagnetic field was also quantized. We worked on this and it should have been written up by me but I left Cambridge before I could manage to finish all the calculations. The material eventually got published in pieces over the next few years.

On another occasion, at a seminar, a distinguished experimentalist stated that he had firm evidence that the Σ had spin 3/2. Julian suggested that I study the spin 3/2 formalism that he and Rarita had developed. Later, Julian's next "assistant," George Sudarshan and I published together the results of his and my work on this problem. Happily, further experiments established that the Σ had spin 1/2 and soon with the rapidly growing number of particle states, the fact that a fundamental field theory for spin 3/2 fields had problems didn't seem to be so important. A student working with me about ten years later and who was attending a course at Harvard given by Julian (as did many of the MIT students) decided that the Rarita-Schwinger formalism should be used for the spin 3/2 hadronic states. I told him what I remembered about it. It didn't do any great harm to him since he is presently your Dean, Roberto Peccei.

The outstanding particle physics research that Julian produced in this period was his 1957 *Annals of Physics* article on fundamental theory. It contains the germs of many of the ideas which bore fruit during the following decade. For example the suggestion of the two neutrino theory, the remark that vacuum expectations of scalar fields could be a way of breaking symmetries and also giving Fermi fields masses, and the introduction of "hypercharge" as a more convenient way of classifying the hadronic states. Finally one shouldn't forget his pioneering attempt to relate the weak and electromagnetic interactions, which his student Sheldon Glashow later succeeded with by postulating in addition neutral current weak interactions.

I cannot finish revealing my memories of that time without also telling you a little about how my wife Gladys and I came to have the great privilege of becoming the friends of Clarice and Julian. We first met Clarice on being invited numerous times to wonderful parties and dinners at their apartment on Fayerweather Street in Cambridge. These occasions took place quite often as Julian was frequently visited by physicists of all varieties and often this became an occasion for an evening celebration. We also came to know Clarice's mother Sadie Carrol who in time we also came to think of as a dear friend. These events would often last until the small hours since Julian was well known as a night owl. Resident at Fayerweather Street there was also a Cat named Leo (I believe, named after Galileo) who indicated when it was time to leave by stationing himself by the front door. Clarice was the kind and gracious hostess at these events and I am sure that everyone who was fortunate enough to know Clarice and Julian remembers those gatherings with great pleasure.

After I left Harvard for a year in Copenhagen, we met Clarice and Julian in Vienna as fellow tourists. Gladys and I often recall those few days with pleasure. We were sad when about 15 years later they left the Boston area for their new life here in the kinder climate of Southern California. Our loss was your gain. I know that you must have taken as much pleasure as we did in knowing both Clarice and Julian. Julian was a truly monumental figure in the theoretical physics of this century and his contributions will long be remembered in the history of our science.

JULIAN SCHWINGER — PERSONAL RECOLLECTIONS

Stanley Deser

Brandeis University

Dear Friends of Julian, Dear Clarice,

Julian Schwinger was a great scientist and a complicated — therefore interesting — human being. It seems such a short time ago that we celebrated his 60th and 70th birthdays here; likewise for the 45 years ago that I first saw Julian, and the 39 years since Elsbeth and I became real friends with Clarice and him, in Copenhagen. It was only a very few months ago that he sent me (via Clarice of course) what was to be his last, kind, message. During those years a lot of memories have accumulated for me, as they have for many of you. Indeed, several of my older fellow-alumni, notably Bryce DeWitt, Abe Klein and Walter Kohn have given their recollections at another recent memorial occasion. Doubtless there will be many more. Our collective memory will thereby help to perpetuate Julian's memory and that will serve as some consolation to us all.

There are a lot of clichés about Julian: he was shy, and not given to small talk; he was hard to approach and did not correspond or write recommendations very readily; he kept odd hours. All that is correct and all beside the point. What is true and relevant is that

Julian was a class act: a true gentleman and a very generous one, as I have witnessed from Cambridge to Belmont to Bel Air (he must have been one of the few inhabitants of the latter who never talked about money or real estate!). Telling examples were his constant heroic and usually successful attempts to deplete his wine collection in favor of his visitors; yet he was understated but happy about his (part) ownership of a high-quality (naturally) vineyard. I also remember the one inverse case: on that October morning in 1965 when the Stockholm call finally came, many of us descended on Belmont with a pretty impressive array of hastily marshalled Champagne, none of which survived the day. Characteristically, Julian never expressed disappointment at that phone call's ten-year delay. [His Nobel lecture was of course up to his usual standards and still makes great reading.] Like all of us, Julian hated to do chores that kept him away from physics, but he did them. He gave his students the support they needed to start their careers, including writing letters when they were really necessary; I know this was true for others, as it was for me, whom he rescued from Europe after a postdoc stretch there had left me jobless. Correspondingly, Julian was reluctant to burden others; I will never forget his asking me — with great embarrassment — to write a letter for him (the Guggenheim Foundation in its impartial majesty requires letters even for Nobel Laureates).

Also not so well appreciated were his broad culture (which he never displayed for effect), his sense of humor and his commitment to many good causes where his name and contribution would help. Many of us could, I know, recite countless examples of all these traits. Likewise, I can attest that his shyness — or rather polite reserve — did not preclude uninhibited conversations on all sorts of topics. Of course, in physics discussions he was usually better and faster, but always tactful and never tried to exert authority; he was always willing to take one's point when he felt it to be valid. Indeed, like most great physicists he was happy to welcome his students as colleagues and peers in our common pursuit.

Recalling the deeds of our heroes is a time-honored way to instruct and uplift the tribe. First, let me belabor the obvious: For many years Julian defined the mainstream of our frontier — he was the source of ideas, methods and directions. I recall a 1950's paper in the "Journal of Jocular Physics" at the Niels Bohr Institute, purporting to be a sure-fire template for writing successful papers, guaranteed to pass any referee. It began with something like "According to Schwinger," continued with "as conveyed by the Green's function expression" and so on — only a few blanks were left to be filled in from one's favorite work by Julian. What's more, we all know that many a serious paper along these lines is indeed to be found in the literature! In addition to QED, it was nuclear physics, formal field theory, classical electrodynamics, particle phenomenology, electroweak interactions, anomalies, monopoles, Schwinger terms and Schwinger models — and much more — that associated Julian's name to the progress of physics in those miraculous years. I will turn to yet another area, gravity and higher spins, in a minute.

Given this remarkable record in one of the more remarkable eras of physics progress, why did Julian leave the mainstream for other, less followed, paths? This question merits a great deal of thought; Moses not entering the Promised Land is the Biblical prototype, but Einstein and Dirac, to mention two of Julian's own heroes, provide two more recent and apposite examples. For Einstein it was the overwhelming force of one overarching principle; for Dirac, a requirement of perfection in his offspring, relativistic quantum field theory. Julian's move must not have been undertaken lightly: he kept the field faith even during the darkest days of excessive dispersion — and even there he advanced its serious side, for example actively inspiring the work that Gilbert, Sudarshan and I were doing. Whatever his reasons, there are lessons for us here: First, mainstreams are not to be left lightly, but they are not sacrosanct either. Second, even our heroes are subject to the stresses that come with the territory of our exciting but difficult frontier. Finally, when leaving the mainstream, be sure to continue to do beautiful and important non-controversial work that maintains your credentials; for Julian it was an impressive array, perhaps best exemplified by his atomic physics studies.

Since the existence of all those generations of his students is proof enough of his efforts and successes, and so much has been eloquently said about his courses and lectures, I shouldn't take your time on this subject, but cannot resist a few remarks. To do new physics requires preparation and Julian made sure his students, and through them future generations, learned much more than Landau's "theoretical minimum." It was total immersion every semester — what enormous labor he must have put in "the night before" all those years — odd hours, indeed! I only wish Julian had completed the E & M textbook he started some time ago, and from which I had the chance to teach. In later years, Julian did not disdain popularization and enjoyed enormously his book and BBC programs on relativity, without succumbing to the temptation of becoming a media personality.

Let me finish with some brief physics comments on one particular stream of Julian's interests that is less widely thought of. It concerns gravity and higher spins, a subject of special interest to me. When he began physics, relativity was, to put it mildly, not front-line; indeed, it was the only basic subject he never taught, in my time at least, even though I know he had read Pauli's article at an early age. His interest in higher spin fields, and the discernment that there is a qualitative change beyond spin 1 began early. In 1941 Schwinger and Rarita (the latter was, coincidentally, the first "live" physicist I met) wrote a short paper pointing to the modern formulation of higher spin fermions, in particular of spin 3/2 and of its invariances when massless. One sees here characteristically a (well-placed) confidence that they are improving on the original Pauli-Fierz work and a connection with possible experimental applications. However, the first time to my knowledge that Julian worked seriously on gravity was in 1958, when Dick Arnowitt and I were embarking on our own spin 2 project; this was right after the dispersion-theoretic work I have already mentioned, and I think we were all reacting to the excesses in that bandwagon with a retreat to field theoretical high (spin) ground! Although my memory of the details is hazy, and Julian did not actually publish anything at the time, I do remember that he too had begun to analyze the free massless spin 2 field and that he talked to me about it (I was his assistant that year). We duly acknowledged this in our linearized gravity paper.

If you look at Julian's gravity-related papers of the early 60's, the gravitational field is first — and mainly — used as a tool to develop the famous stress tensor commutation relations required for consistency of a field theory; these deep ideas were also being elaborated by Dirac. [Curiously, though, there is no mention of Schwinger terms, which he had invented in the current-current commutator context.] Here we see explicit statements of the difference between lower and higher spins, as well as a mastery of matter-gravity coupling technology in modern vierbein form. In a later paper comes the application to gravity as a dynamical system (rather than as an external field) formulated in terms of a set of canonical variables to those Arnowitt, Misner and I had developed. Julian kept his strong interest in gravitation after he created source theory, in terms of which he treated e.g., relativistic corrections to Newtonian motion. Long afterwards, when supergravity was discovered, Julian became very interested, and no wonder: Here were his old friends spin 3/2, Grassmann variables, deep gauge invariances and gravity all mixed together. Julian decided on total immersion learning for himself: I was summoned across the country to the physics seminar room at UCLA to give him a private weekend tutorial, on the all-day Russian scale; it did not surprise me that Julian was a fast learner! People have wondered why he, who was so well placed for it, never discovered supersymmetry himself; I think that (apart from the fact that you can't win 'em all) it was his anchor in experiment that kept him from this type of unification at the time. Yet, as in other fields, he had uncannily laid the groundwork for these new ideas. It is for such reasons that we all instinctively miss the passing of our giants, even if they were not personal friends: we need their shoulders to stand on.

This foreshortened set of vignettes is of course only a caricature of the measure of the man, but I hope it has found some resonance with your own recollections. Julian's place in both the physics Pantheon and in our hearts is secure!

JULIAN SCHWINGER: THE QED PERIOD AT MICHIGAN AND THE SOURCE THEORY PERIOD AT UCLA

Robert Finkelstein

University of California, Los Angeles

Although I did not take my degree with Julian, I — like so many others — must count myself as one of his students, first because of his immense influence on the physics of our times, then in my own case because I attended his Michigan lectures of 1948, and most importantly as a result of having enjoyed his friendship for many years after he came to UCLA. I shall say a few words first about the early QED period at Michigan and then about the later source theory period at UCLA.

It was in the 1948 Michigan lectures that Julian first described his breakthrough in quantum electrodynamics to a wider audience. That was almost 50 years ago. Among the young people present were Dyson, Kroll, Lee and Yang. (Yang told me that he had never heard anyone speak English so rapidly!) One may give a feeling for the impact of these lectures by quoting Dyson who wrote home that "in a few months we shall have forgotten what pre-Schwinger physics was like." The work Julian was then describing grew out of the experimental discoveries of Lamb, Rabi, and Kusch and led to the mid-century revolution in theoretical physics. Bethe at that time described this period as the most exciting in physics since the great days of 1925-30 when quantum mechanics was being discovered.

Although very many others, and of course particularly Tomonaga and Feynman, contributed to this development it was Julian who made the major breakthrough by first understanding the full consequences of the new experiments, by constructing the first manifestly covariant theory, and by first calculating in lowest order all the previously inaccessible consequences of QED. Other simpler formalism soon followed: Feynman's Michigan lectures came the following summer, and Dyson's lectures came the third year, but a special place in our Pantheon belongs to Julian who first climbed the mountain and dominated the earliest developments. To show that this mountain could be climbed at all was a very great achievement because up to that time quantum electrodynamics had appeared to be fatally flawed.

The Schwinger theory of 1948, while adequate for its original purpose, was, like every first invention, relatively crude and could not be pushed to higher order in any simple way. Therefore during the fifties he developed increasingly powerful calculational techniques. To this period belong the Schwinger action principle, the uniform treatment of bosons and fermions, many-particle Green's functions, and the extensive use of functional techniques.

During the fifties, it also became apparent that to go on, new physical ideas and/or new experiments were needed. As a consequence there were new theoretical developments such as dispersion theory, S-matrix theory, axiomatic field theory, even a C* algebra group, and in the background of all these programs was the infinity problem which was continually emphasized by Dirac. Julian's own major response did not come until the sixties when he transferred his attention from the Green's functions to their sources. The source theory work was begun at Harvard but most of the later development was carried out at UCLA.

Source theory represented Julian's effort to replace the prevailing operator field theory to which he had contributed so richly and so fundamentally by a philosophy and methodology that eliminated all infinite quantities. He did in fact succeed in constructing an infinity-free formalism that was also friendly to the introduction of new experimental information and new theoretical ideas. Moreover it was not simply a program: he and his UCLA source group, Kim Milton, Lester de Raad, Wu-Yang Tsai, made many applications to high energy physics and showed that it was a very effective calculational tool. Since source theory, like every successful physical theory, necessarily shared features with the formalism it was replacing, some felt it was nothing really new — but of course it was.

In comparing operator field theory with source theory Julian revealed his political orientation when he described operator field theory as a trickle down theory (after a failed economic theory) — since it descends from implicit assumptions about unknown phenomena at inaccessible and very high energies to make predictions at lower energies. Source theory on the other hand he described as anabatic (as in Xenophon's Anabasis) by which he meant that it began with solid knowledge about known phenomena at accessible energies to make predictions about physical phenomena at higher energies. Although source theory was new, it did not represent a complete break with the past but rather was a natural evolution of Julian's work with operator Green's functions. His trilogy on source theory is not only a stunning display of Julian's power as an analyst but it is also totally in the spirit of the modest scientific goals he had set in his QED work and which had guided him earlier as a nuclear phenomenologist.

At the same time his conservative approach did not prevent him from anticipating the electroweak theory and inspiring Shelly Glashow, or from prefacing this electroweak paper with the quotation from Einstein: "The axiomatic basis of theoretical physics cannot be extracted from experience but must be freely invented." It also did not prevent him from proposing a dyon theory of matter. Yet his own speculations as well as his interest in contemporary work, were always, without exception, strictly ruled by his scientific integrity. His basic scientific conservatism was moreover totally consistent with his antipathy toward the arrogant dismissal of unpopular ideas.

Although Julian was continually productive there are, in addition to the source theory program, only two other UCLA projects which I would like to mention today.

The first was an extensive study of multi-electron atoms carried out in collaboration with B.G. Englert. This work developed out of an undergraduate course in quantum mechanics and illustrates the fact that every Schwinger course was partly a research project. One of the sidelights of this work came to be known in the mathematical community as the Schwinger conjecture. It has recently been proved in a 186-page paper by the mathematicians Fefferman and Seco.

The second project was stimulated by Seth Putterman's experiments on sonoluminescence. In these experiments a single oscillating bubble of gas in water absorbs sonic energy and emits light. Surprisingly, the spectrum of the light extends into the ultraviolet and perhaps more surprisingly the most copious sources of the light are the inert gases. Julian speculated that the light is dynamic Casimir radiation and studied this radiation by — of course! — source techniques. The problem posed by this speculation is genuinely complicated since there are interacting hydrodynamic and electromagnetic effects. The analysis had not been completed at the time of Julian's death, but he had worked on this problem up until his fatal illness.

On the occasion of Julian's 60th birthday, Rabi gave some fatherly advice to physicists who had reached the age of 60. He cautioned them not to compete with the kids. I think that advice was not necessary for Julian since I believe that in one sense he never competed with anyone. He simply set his own scientific goals and forged ahead. Although he of course knew his powers and he knew what he had accomplished, in a deep sense his modesty was genuine. Julian was not only one of the greatest physicists of our times but he was a beautiful example of an unpretentious, gentle, and friendly man in an intensely rough and tumble profession. Today we honor not only a great man but also a rare individual and for many of us a dear friend.

Clarice asked me shortly after Julian's totally unexpected illness and sudden death: "Does all this seem real?" It certainly didn't then and I'm afraid it doesn't even now.

Yet — to borrow the words he himself spoke at the memorial for Tomonaga, Julian "lives on in the minds and hearts of the many people whose lives he touched and graced"; or as he said of George Green "he is, in a manner of speaking, alive, well, and living among us."

JULIAN SCHWINGER MEMORIAL TRIBUTE

David Saxon

Concluding Remarks

Earlier, you will recall, I quoted Sam Goudsmit's remarks about Julian, the Nobel Prize and propitious times. In large part because of wartime work on microwaves, work going on even as Goudsmit spoke, the period following the War unexpectedly turned out to be "as exciting as the great days of the 1920's," as Bethe put it. Sam was right. When the time was ripe, Julian at once carried out the work which earned him his Nobel Prize.

Julian Schwinger was a genuine prodigy and a great man. His death is a terrible loss to the world of physics and to those of us who were his students, collaborators and friends, to those of us who were close to him, and especially to Clarice, his devoted wife of 47 years. Clarice, you have our affection, support and deepest sympathy.

Over these past three months, thoughts of Julian have dominated my mind: who he was, what he was, what he meant to me. But I have been irresistibly drawn to the larger question of what his death means to mankind. I can put it in a sentence. Julian was one of those rare souls whose existence serves to remind and reassure us that we are more than animals. So was Mozart. Music is a universal language accessible to all; the language of physics is not. Those of us who have lived with physics are privileged in our unique understanding of what the world has lost.

Schwinger with R. Feynman at the Lepton-Photon Symposium at Stanford in 1975. J. Bjorken is in the background.

Schwinger with some of his doctoral students and W. Rarita (second row, far left) at UCLA on his 60th birthday. Flanking him are M. Kivelson, Schwinger's only female student, and the editor.

Presenting relativity to a general audience on BBC TV. (Courtesy of BBC/Open University.)

SCHWINGING A SORCERER'S WAND: JULIAN AND I

Y. Jack Ng

University of North Carolina at Chapel Hill

I was among Prof. Schwinger's last students, probably the last one from Harvard. For a couple of years I was virtually his only thesis student, and so I had some rare opportunities to interact and socialize with him. I need not dwell on Schwinger's prowess as a physicist, since that is well known. I will instead tell some Schwinger stories — small stories that also involve me. (More than once, Schwinger warned me not to believe every story told about him. But these stories are true.)

My first meeting with him in his office at Harvard is one that I have always recalled with fondness. A couple of weeks ago he had assigned me a small problem following my preliminary exam, and I was briefing him on the progress I had made. He thought for a while and then suggested a certain term I should consider in my calculation. Immediately I told him that there could not be such a term because of parity conservation. Impressed with my speedy (and correct) rebuttal, he hit himself on the head and said repeatedly, "How stupid I am!" Fancy a lowly graduate student hearing such a humble self-deprecatory remark from a giant of modern physics! My research career had started rather auspiciously.

Of course, Schwinger was demanding not only of himself, but of his students as well. When I asked him to take me on as his student, I made it a point to tell him I worked very hard. "That is the least I expect from my students," was his terse reply. Throughout my years with him I never saw him become upset — except for the one time when he was mildly angry with himself and with me for our failure to notice a subtle infrared structure in a bound state problem. I believe he must have been fairly disappointed with himself for not coming up with the idea of supersymmetry.

I was the only graduate student from Harvard who accompanied Schwinger to UCLA in spring 1971. (At that time I had no inkling that he was going to stay in Los Angeles.) My wife Woonyu and I thus became acquainted with the Schwingers considerably better than most of his earlier students did. We saved their wedding gift to us (a pair of kitchen knives) and their gift years later to our older son (a silver spoon) as souvenirs. During my student years the Schwingers invited us to their parties three times, and in return they came to our house twice for barbecues. In one of the social gatherings in their house, someone asked if there had ever been two US presidents who were born or died on the same day. I cited the example of Jefferson passing away a few hours before the elder Adams. Incredulous, Schwinger promptly left to consult his *Encyclopedia Britannica*. (Anyone who has read his Nottingham lecture in memory of George Green, wittily entitled "The Greening of Quantum Field Theory: George and I," knows that Schwinger loved to read the *Encyclopedia Britannica*.) For the second and last time in my life I had told the master something he did not know offhand.

I defended my Harvard Ph.D. thesis in the Schwingers' living room in Bel Air, CA. Mrs. Schwinger teasingly asked me if I wanted her to hold my hand during the exam to keep me calm. Noticing that some glasses of champagne were already prepared nearby, I smiled and confidently declined her offer. As expected, the exam was a breeze. Then it was time for me to face the outside world. Even after we had left UCLA, we kept in touch with the Schwingers, at least once a year, by exchanging Christmas cards (although in recent years, the correspondence has become one-way, from us to them). I learned that the Schwingers actually enjoyed our annual letters and cards. I was at UCLA to attend the Schwinger symposia on the occasions of his 60th and 70th birthdays. The last conversation I had with the master was to ask for help in solving the non-perturbative Schwinger-Dyson equation in the presence of strong fields. He listened carefully, but at the end he could offer no help. Then like a father he gently reminded me, "You are on your own now, kid." The last thing I ever heard Schwinger say was addressed softly to John Bell right after Bell's talk in the second Schwinger symposium: "I hope you don't mind if I disagree?"

It is common knowledge that Prof. Schwinger's lectures were smooth as silk. In lectures, as in his research, he strove for perfection. Like a virtuoso soloist in a concert, he wanted to perform flawlessly and brilliantly in every lecture. And like a sorcerer casually swinging his wand (or was it chalk?) he conjured up magic on the blackboard in front of our eyes.

Some people claim that Schwinger's downfall was precipitated by his conception of source theory. As one trained by Schwinger in that art, I want to defend his point of view, as I once did in a seminar at SLAC which Sid Drell dubbed "Sorcery by the Sorcerer's Apprentice." Source theory is an approach that combines the salient features of operator field theory and S-matrix theory. And if nothing else, it provides conceptual clarity and tremendous calculational power. (For the skeptics, try calculating the helicity cross-sections for electron-electron scattering to one-loop.) Schwinger was obviously disappointed with the rather cold reception from the physics community to the source theory approach. Alas, as Schwinger probably realized, 'tHooft's proof of the renormalizability of gauge theories considerably diminished the appeal of his source theory, a more phenomenologically oriented approach, at least in the study of strong and electroweak interactions.

It is ironic that Schwinger, who played such an important role in the development of operator field theory, found himself rejecting it in the later part of his life. He stuck staunchly to his source theory approach to the very end. Some would charge him of stubbornness. Curiously, I think he would have gladly pled guilty to that. "Stubborn? Who isn't?" he used to ask me. As one who changed the landscape of modern physics, he was not afraid to go against the tide of fashion or belief in our field. Nor was he afraid to espouse different points of view or to champion different ways of thinking and doing physics. He prefaced his set of books *Particles, Sources, and Fields*, with the quote, "If you can't join 'em, beat 'em." But always physics was his love. Perhaps Robert Frost's epitaph could equally well serve as his:

"I had a lover's quarrel with the world."

With the passing of Julian Schwinger, we have lost one of the foremost sorcerers of all time in the physics community, one whose lectures charmed and inspired us, and one whose work bewitched and stimulated us all. He opened our eyes to deep truths and great beauties.

JULIAN SCHWINGER — REMINISCENCES AND NUCLEAR PHYSICS*

H. Feshbach

Massachusetts Institute of Technology

I first heard of Julian Schwinger some 60 years ago. My physics instructor at CCNY, Lloyd Motz, since that time a professor at Columbia University, told me of an extraordinary brilliant and young physicist. Actually at that time Motz and Schwinger published a paper "On the β-Radioactivity of Neutrons." It was one of two papers by Schwinger published during 1935. The other was with Otto Halpern, "On the Polarization of Electrons by Double Scattering." He was only 17 years old. In 1937 at the advanced age of 19 he published seven papers.

I recall from those days a number of vignettes. Schwinger giving a lecture on general relativity to the CCNY math club; Schwinger giving a colloquium at Columbia on the 1934 Fermi theory of β decay. Schwinger and a number of us including Morton Hamermesh and Max Schiffman (a student of Courant's) listening to music at Artie Levinson's apartment. We were a young New York group, devoted to

---

*Presented at the Washington meeting, April 21, 1995 of the American Physical Society.

physics, attending Courant's courses at NYU and a seminar directed by Rabi and Gregory Breit. It was a very exciting time.

To appreciate Schwinger's contributions to nuclear physics bear in mind the situation in that field at that time. Remember that one knows the existence of the deuterons but one doesn't know if it is the only bound state of the $n - p$ system. Although the spins of the deuteron and proton are known, the spin of the neutron is not. The $n - p$ scattering cross-section at low energies is thought to be known (the value at the time is incorrect) but not its decomposition into singlet and triplet amplitudes (Or triplet and quartet if the neutron spin is 3/2). Although rough values of the range and shape of the nucleon-nucleon potential are known, a detailed knowledge of the structure of this potential is still to be discovered. Charge independence had been proposed but it is not yet proven. The quadrupole moment and the magnetic moment of the deuteron are yet to be measured. Turning to post war discoveries, the analysis of "high energy" nucleon-nucleon scattering demonstrated the presence of a short range effectively repulsive component of nuclear forces, the so-called hard core. A start on a shell model had been made by Feenberg and Wigner but its eventual description was post war as was the discovery of the rotational model. Turning to reaction dynamics, compound nuclear resonances had been observed in the early 1930's, the statistical theory of nuclear reactions was proposed in 1937. The direct reactions processes as well as their contribution to the delineation of the shell model were yet to be discovered. In summary most of the present day understanding of nuclear structure, nuclear reactions, and nuclear forces was still to be developed.

His papers of 1937 and 1939 can be divided into two groups. The close connection in this period of his research with experiment is characteristic. One set was devoted to fundamental questions regarding the nature of nuclear forces. Their spin dependence was the focus of Schwinger's paper with Teller entitled "The Scattering of Neutrons by Ortho and Para Hydrogen." The spin of the protons in ortho-hydrogen are parallel while in para-hydrogen they are anti-parallel. The scattering of neutrons by these two molecules will differ because the scattering amplitude will involve differing and interfering combinations of the singlet and triplet scattering amplitudes. It was thus in principle possible to obtain these amplitudes by comparing the scattering from the two molecular types. Schwinger in 1940 pointed out that this experiment would also determine the spin of the neutron. But there were important complications. The effects of inelastic processes and the errors because of the use of the "Fermi Approximation" had to be calculated. This was the focus of a number of papers with Morton Hamermesh in 1939, 1946 and 1949. It was also a topic in his famous paper with Lippman entitled "Variational Principles for Scattering Processes" (1950). In this paper the Lippman-Schwinger equation which had a enormous influence on scattering and reaction theory was proposed.

The discovery of the quadrupole moment of the deuteron led to the influential series of papers written with Rarita and published in 1941. In these papers the full consequences for the low energy $n - p$ system of the tensor force required by the deuteron quadrupole moment and magnetic moment were carried out. These are classic papers not only because of the physics that is so clearly developed but also because of the mathematical techniques he developed. I later worked with him on the same subject (1951). It was during this year (1951) that he developed a way for dealing with angular momentum algebra.

In a second set of researches in the 1937 period he worked with the experimentalists H. H. Goldsmith and John Manley. His role was the interpretation of their experiments which involved the use of neutron induced resonance reactions. One of the important consequences of this work was the correct determination of the neutron-proton cross-section. The previously accepted value was off by about 50%.

Schwinger left the New York City area sometime around 1939–1940 eventually ending up in Berkeley where he wrote two papers with J. Robert Oppenheimer, one of which revealed the possibility of the radioactive decay by means of pair emission. I didn't see him again till around 1943. I met him by chance at MIT and of course was delighted to see him and discovered that he was at the MIT Radiation Laboratory. I immediately invited him to dinner. He asked, "when do you eat dinner?" Oh, I said about 7PM. "Well," he replied, "that's when I have my breakfast." He came anyway!

Schwinger worked at night. The story is told (and I have verified it) that he often would examine the calculations of his colleagues evaluating their integrals or leaving useful references. It was during this period at the Radiation Laboratory that he developed a complete theory of electromagnetic wave guides as well as developing new methods for dealing with apertures. Papers with Harold Levine and N. Marcuvitz were published after the war. A book with Marcuvitz was projected but only a paper in 1951 was published. As these papers show, Schwinger had an awesome ability (I should say unequaled) to deal with problems in the classical mathematics generated by problems in physics.

After the war, Schwinger came to Harvard. It was an extraordinary time. The students were superb and Schwinger had many. We ran a most delightful seminar, on Thursday evenings if I remember correctly. We often had dinner first at a French restaurant in Harvard Square. I remember one occasion when we were talking physics at such a dinner when he announced that he had just worked out the polarization that would be induced in neutrons upon scattering by ^4He. But then he went on to add that he hadn't slept for the last 36 hours working the problem out. This was also the period of the conference at Shelter Island, at a resort in the Poconos and finally at Oldstone on the Hudson. Weisskopf, Schwinger and I went down to the New York headquarters of the APS by train where we were joined by the rest of the group going to Shelter Island. There we were taken by bus with a police escort all the way to Greenport where one took the ferry to Shelter Island. The Pocono conference was where both Schwinger and Feynman spoke. Schwinger's lectures were always, whether it was in the classroom, at seminars, or at the Pocono conference, carefully prepared and superbly delivered. After the Pocono conference we drove back to Cambridge together. I remember his comment that he was amazed that Feynman obtained identical results to his although the methods appeared to be so different. We also talked about Dirac's presentation at the conference of a theory of magnetic poles. But we didn't make much progress. On the way we had a flat. I changed the guilty tire. But to start the car again, I had to lift the hood and join two wires in order to start the car! Schwinger had a new car very shortly thereafter.

Most of the physics graduate students and a fair fraction of the faculty in the Cambridge area attended Schwinger's course in nuclear physics. Nuclear physics was fortunate because the lectures were written up by John Blatt and made available to a wide audience. These notes form an excellent introduction to the applications of quantum mechanics in which a number of elegant methods of wide applications were developed. The concept of effective range, the consequences of the existence of non-central nuclear forces, the novel use of variational methods for scattering problems, the interaction of nuclear systems with electromagnetic fields etc. were discussed. One cannot exaggerate the impact of these notes on the physics graduate students of the late forties and fifties by which time they had become a general part of the theorist's armamentarium.

In the late forties Schwinger and I embarked on the calculation of the properties of the $n - p$ system with special emphasis on the deuteron. These calculations, now of historical value only, was one of the first to use a computing machine located at Harvard. You will notice that the results were announced in 1948 but not published until 1950. The circumstances were as follows. I wrote the paper — which was unusual since Schwinger usually did the writing when he collaborated. We then went over my draft literally word for word. At the rate of one meeting a week it took a year to obtain the final version. It was an education for me.

There were few contributions to nuclear physics once field theory became the focus of Schwinger's research. The use of neutron or proton scattering by ^4He as a polarizer as a consequence $p_{1/2}$ and a $p_{3/2}$ non degenerate resonances. But of course it can also be used as an analyzer and continues to be used in many experiments. Another contribution is referred to as Schwinger polarization scattering, a consequence of the interaction of the neutron magnetic moment with the electric field of the nucleus. Schwinger showed that neutron polarization of 100% can be achieved by small angle scattering of neutrons by nuclei.

Another set of papers which should be mentioned, though strictly they are not nuclear physics, are those which have to do with synchrotron radiation. Schwinger discovered synchroton radiation. These papers begin with the paper in 1946 on electron orbits in the Synchrotron with D. Saxon and continued with additional papers in 1946, 1949, 1954, 1973, 1974 and 1976.

My personal interactions with Schwinger and his wife Clarice continued for many years, although they were considerably reduced when they moved to UCLA. In 1954 I took a "subway Guggenheim" with Schwinger working principally with Paul Martin. I, of course, went to Schwinger's lecture. We had lunch together on Schwinger's lecture days with Paul and Margaret Kivelson and others. Clarice Schwinger inaugurated a yearly dinner which included Viki Weisskopf, Bernie Feld and myself. The ostensible excuse was that Schwinger and I had birthdays early in February. He is merely one year younger. I learned then of his interest in music and in English literature. The Nobel award was of course a great event. We all assembled in his Belmont home to celebrate. Needless to say his departure to California was a great shock but we maintained contact although I am a very poor communicator. I visited them several times and vice versa when he came to Cambridge. We played tennis. We were both students of Asim Yildiz. We talked physics and we participated in each other's birthday parties at the age of 60 and 70. The last time I saw Julian was at a celebration of Ken Johnson's sixtieth birthday held at MIT.

Julian Schwinger was my dear friend for nearly sixty years. He was a great and unique physicist. But he was also a wonderful person. He was gentle and I never heard him speak sharply to anyone — even if the seminar speaker was being painfully obscure or just plain wrong. I was very fond of him. I liked him and I miss him.

Papers Referred to in the Text

- On the polarization of electrons by double scattering (with O. Halpern), *Phys. Rev.* **48**, 109 (1935).
- On the β-radioactivity of neutrons (with Lloyd Motz) *Phys. Rev.* **48**, 704 (1935).
- On the magnetic scattering of neutrons, *Phys. Rev.* **51**, 648 (1937).
- On non adiabate processes in inhomogeneous fields, *Phys. Rev.* **51**, 544 (1937).
- The scattering of neutrons by ortho and para hydrogen (with E. Teller), *Phys. Rev.* **51**, 775 (1937), **52**, 286 (1937).
- Depolarization by neutron-proton scattering (with I. I. Rabi), *Phys. Rev.* **51**, 1003 (1937).
- On the spin of the neutron, *Phys. Rev.* **52**, 1250 (1937).
- The scattering of neutrons by hydrogen and deuterium molecules, *Phys. Rev.* **55**, 679 (1939) with M. Hamermesh.
- Neutron scattering in ortho and para hydrogen and the range of nuclear forces, *Phys. Rev.* **58**, 1004 (1940).
- The scattering of slow neutrons by ortho and para deuterium with M. Hamermesh, *Phys. Rev.* **69**, 145 (1946).
- Neutron scattering on ortho and para hydrogen with M. Hamermesh, *Phys. Rev.* **71**, 678 (1947).
- Neutron energy levels (with J. Manley and H. Goldsmith), *Phys. Rev.* **51**, 1022 (1937).

- The widths of nuclear energy levels (with J. Manley and H. Goldsmith), *Phys. Rev.* **55**, 39 (1939).
- The neutron-proton cross section (with V. Cohen and H. Goldsmith), *Phys. Rev.* **55**, 106 (1939).
- The resonance absorption of slow neutrons in indium (with J. Manley and H. Goldsmith), *Phys. Rev.* **55**, 106 (1939).
- On the pair emission in the proton bombardment of fluorine (with J. R. Oppenheimer), *Phys. Rev.* **56**, 1066 (1939).
- On the interactions of mesotrons and nuclei (with J. R. Oppenheimer), *Phys. Rev.* **60**, 61 (1941).
- The photodisintegration of the deuteron (with W. Rarita and H. Nye), *Phys. Rev.* **59**, 209 (1941).
- On the neutron-proton interaction (with W. Rarita), *Phys. Rev.* **54**, 436 (1941).
- On the exchange properties of the neutron-proton interaction (with W. Rarita), *Phys. Rev.* **59**, 556 (1941).
- On a theory of particles with half integral spin (with W. Rarita), *Phys. Rev.* **60**, 61 (1941).
- The quadrupole moment of the deuteron and the range of nuclear force, *Phys. Rev.* **60**, 164 (1941).
- On tensor forces and the theory of light nuclei (with E. Gerjuoy), *Phys. Rev.* **61**, 138 (1942).
- On a phenomenological neutron-proton interaction (with H. Feshbach), *Phys. Rev.* **84**, 194 (1951).
- On the theory of diffraction by an aperture in an infinite plane screen (with H. Levine), *Phys. Rev.* **75**, 1423 (1949).
- On the radiation of sound from an unflanged circular pipe (with H. Levine), *Phys. Rev.* **72**, 742 (1947), **73**, 383 (1948).
- A variational principle for scattering problems, *Phys. Rev.* **72**, 742 (1947) (with H. Levine), *Phys. Rev.* **74**, 1212 (1948).
- On the transmission coefficient of a circular aperture (with H. Levine), *Phys. Rev.* **75**, 1608 (1949).
- Variational principles for scattering processes I. (with B. Lippman), *Phys. Rev.* **79**, 469 (1950).
- On the representation of the electric and magnetic fields produced by currents and discontinuities in wave guides (with N. Marcuvitz).
- Radiation forces and torque (with H. Levine), *Phys. Rev.* **87**, 224 (1952).
- *Discontinuities in Wave Guides* (with D. Saxon), 1968.
- On the polarization of neutrons by resonance scattering in helium, *Phys. Rev.* **69**, 681 (1946).
- On the polarization of fast neutrons, *Phys. Rev.* **73**, 407 (1948).

- Electron orbits in the synchroton (with D. Saxon), *Phys. Rev.* **69**, 702 (1946).
- Electron radiation in high energy accelerators, *Phys. Rev.* **70**, 798 (1946).
- On the classical radiation of accelerated electron, *Phys. Rev.* **75**, 1912 (1949).
- The quantum correction with radiation by energetic accelerated electrons, *Proc. Nat. Acad. Sci.* **40**, 132 (1954).
- Classical radiation of accelerated electrons II, a quantum viewpoint, *Phys. Rev.* **D7**, 1696 (1973).
- Radiative polarization of electrons (with W. Tsai), *Phys. Rev.* **D9**, 1843 (1974).
- Classical and quantum theory of synergic synchrotron-cerenkov radiation (with W. Y. Tsai and T. Erka), *Ann. Phys.* **96**, 303 (1976).
- New approach to quantum corrections in synchrotion radiation, *Ann. Phys.* **110**, 63 (1978).

AN IMPORTANT SCHWINGER LEGACY: THEORETICAL TOOLS

Lowell S. Brown

*Department of Physics, University of Washington,
Seattle, Washington 98195*

Abstract

This article is a transcription of a talk given at the April 1995 Meeting of the American Physical Society in Washington DC. The talk showed a number of transparencies which reproduced some of Schwinger's published work as well as other relevant journal pages. Some of these transparencies have been incorporated into the text; other transparencies are displayed here as 'figures'.

Today I'd like to talk about what I consider to be a very important legacy of Julian Schwinger: His development of tools for theoretical physics. The importance of the development of tools for experimental methods is certainly well recognized. To my mind, the development of tools to examine the consequence of physical theory is of equal importance, and so today I'd like to commemorate the memory of Julian Schwinger by giving a brief account of some of the theoretical tools that he either invented or developed.

The utility of Schwinger's methods have long been appreciated. Inside the *Scientific American* of March of 1956 appeared the standard form shown in Fig. 1. All you had to do was to fill in your title

Standard form FT/3.

Title: .... in field theory.

Author: ....

    According to Schwinger

(1)

whence

(2)

hence

(3)

    1)
Thus

(4)

which is not inconsistent with the assumption that

(5)

In virtue hereof

(6)

whence

(7)

hence

(8)

Thus, from a formal point of view

(9)

It is hoped that this kind of argument may lead to the most general formulation of the problem of ghost states.

    The author is .... valuable criticism.

1)
    Cf. also Källen, Ark. f. astr., mat. och fys. ....

STANDARD FORM for a paper in quantum field theory was designed by J. Lindhard and P. Kristensen of the University Institute for Theoretical Physics in Copenhagen. The blanks labeled with the numbers in parentheses at the right are for equations. "Schwinger" is Julian Schwinger of Harvard University, one of the principal workers in quantum field theory.

Fig. 1

and fill in your name, and then you could write: *According to Schwinger,....* Note the conclusion, "It is hoped that this kind of argument may lead to the most general formulation of the problem of ghost states."

Various themes appear throughout Schwinger's work. One of them is to elucidate theoretical structure by a clean notation utilizing operator techniques. This is illustrated in a paper that Herman Feshbach referred to (in the previous talk), "The Variational Principles for Scattering Processes" by Schwinger and Lippmann, and in this paper appears the Lippmann-Schwinger equation. So here it is, here is the abstract operator form of the integral equation that represents the solution to the Schrödinger equation incorporating boundary conditions (and the text that follows):

$$\Psi_a^\pm = \Phi_a + \frac{1}{E_a \pm i\epsilon - H_0} H_1 \Psi_a^\pm. \qquad (1.61)$$

These equations provide a time-independent formulation of the scattering problem, in which the small positive or negative imaginary addition to the energy serves to select, automatically, outgoing or incoming scattering waves.

Another recurring theme that appears in many of Schwinger's papers is a statement equivalent to:

If you don't know it, well..., *vary it*.

I'll give a quote from a six part text book on quantum field theory that appeared in *The Physical Review*. This is Part V of "The Theory of Quantized Fields" which was received on October 26, 1953. It was probably published soon after rather than later, as nowadays.

We thus encounter the differential expression

$$\text{Tr } X^{-1} \delta X = \delta(\log \det X), \qquad (25)$$

which, together with the initial condition, det $1 = 1$, completely defines the determinant of a matrix (or operator).

Let me elaborate on that a bit. Here's the determinant of an infinite matrix. Well, what's the determinant of an infinite matrix? It's not infinitely complicated. All you do is realize that the variation of the logarithm of the determinant of this infinite matrix can be written as the trace of the inverse times the variation of the matrix or the operator — which is still abstract. But you realize that this is just a convolution integral that ordinary people can understand, so then you can have a concrete realization about what an infinite determinant really is:

$$\mathrm{Tr}\, X^{-1}\delta X = \int dx\, dy\, X^{-1}(x,y)\delta X(x,y).$$

Yet another recurrent theme in the work of this period, which is somewhat later than what Feshbach was talking about, is the use of functional methods and effective action. The paper that started it, "On the Green's Functions of Quantized Fields," appeared in the *Proceedings of the National Academy of Sciences* in 1951, and began:

> The temporal development of quantized fields, in its particle aspect, is described by propagation functions, or Green's functions. The construction of these functions for coupled fields is usually considered from the viewpoint of perturbation theory. Although the latter may be resorted to for detailed calculations, it is desirable to avoid founding the formal theory of the Green's functions on the restricted basis provided by the assumption of expandability in powers of the coupling constants. These notes are a preliminary account of the general theory of Green's functions in which the defining property is taken to be the representation of the fields of prescribed sources.

After writing down the field equations for quantum electrodynamics, Schwinger states that:

> With regard to commutation relations, we need only note the anticommutativity of the source spinors with the Dirac field components.

He then introduces the effective action $\mathcal{W}$ and writes:

> the dynamical principle can then be written
>
> $$\delta \mathcal{W} = \int_{\sigma_2}^{\sigma_1} (dx)\langle \delta \mathcal{L}(x) \rangle, \qquad (6)$$
>
> where

$$\langle \delta \mathcal{L}(x) \rangle = \langle \bar{\psi}(x) \rangle \delta \eta(x) + \delta \bar{\eta}(x) \langle \psi(x) \rangle + \langle A_\mu(x) \rangle \delta J_\mu(x). \qquad (7)$$

I'd like to dwell on one paper which is a jewel that still sparkles brightly and which brings forth some other techniques, in particular the method of proper time. I should first note that this paper contains an explicit derivation of the chiral anomaly, the anomaly in the divergence of the axial current in quantum electrodynamics. This is the paper "On Gauge Invariance and Vacuum Polarization" which was published in *The Physical Review* in 1951. Here's the anomaly equation as written in that paper:

$$\partial_\mu [\text{tr } \gamma_5 \gamma_\mu G(x,x)] = -(e^2/2\pi^2) \mathcal{G} \lim_{x' \to x''} \Phi(x', x'')$$

$$= -(2\alpha/\pi)\mathcal{G}, \qquad (5.24)$$

where

$$\mathcal{G} = \frac{1}{4} F_{\mu\nu} F_{\mu\nu}{}^* = \mathbf{E} \cdot \mathbf{H}. \qquad (3.30)$$

was previously defined. The trace on the left of Eq. (5.24) represents the axial current for a Dirac field interacting with an external electromagnetic field, and Schwinger calculates that the divergence of the axial current is F times F dual or $\mathbf{E}$ dot $\mathbf{H}$.

vacuum current vector,

$$\langle j_\mu(x)\rangle = ie\, \mathrm{tr}\gamma_\mu(x|G|x), \qquad (2.20)$$

is obtained from an action integral by variation of $A_\mu(x)$. This is accomplished by exhibiting

$$\delta W^{(1)} = \int (dx)\delta A_\mu(x)\langle j_\mu(x)\rangle = ie\, \mathrm{Tr}\gamma\delta A\, G \qquad (2.21)$$

as a total differential, subject to $\delta A_\mu(x)$ vanishing at infinity. In the second version of $\delta W^{(1)}$, δA_μ denotes the operator with the matrix elements

$$(x|\delta A_\mu|x') = \delta(x-x')\delta A_\mu(x), \qquad (2.22)$$

and Tr indicates the complete diagonal summation, including spinor indices and the continuous space-time coordinates. Now

$$-e\gamma\delta A = \delta(\gamma\Pi+m), \qquad (2.23)$$

and

$$G = \frac{1}{\gamma\Pi+m} = i\int_0^\infty ds\, \exp\{-i(\gamma\Pi+m)s\}, \qquad (2.24)$$

so that

$$ie\, \mathrm{Tr}\gamma\delta A\, G$$

$$= \delta\left[i\int_0^\infty ds\, s^{-1}\, \mathrm{Tr}\, \exp\{-i(\gamma\Pi+m)s\}\right], \qquad (2.25)$$

in virtue of the fundamental property of the trace,

$$\mathrm{Tr}AB = \mathrm{Tr}BA. \qquad (2.26)$$

Thus, to within an additive constant,

$$W^{(1)} = i\int_0^\infty ds\, s^{-1} e^{-ims}\, \mathrm{Tr}\, \exp\{-i\gamma\Pi s\}$$

$$\qquad (2.27)$$

$$= \int (dx)\mathcal{L}^{(1)}(x),$$

where the lagrange function $\mathcal{L}^{(1)}(x)$ is given by

$$\mathcal{L}^{(1)}(x) = i\int_0^\infty ds\, s^{-1} e^{-ims}\, \mathrm{tr}(x|\exp\{-i\gamma\Pi s\}|x). \qquad (2.28)$$

Fig. 2

Figure 2 shows some details in the paper, the details of the proper time method using effective actions. Again you encounter this beautifully clean notation and clever use of operators. The vacuum expectation value of the electromagnetic current is represented as the matrix element of an operator Green's function, and again we vary something. The variation of the effective action is thus the trace of the variation of the field times this operator Green's function. The operator Green's function can be represented by a proper time integral, and using this representation the variational equation may be

> It is now appropriate to notice that the integral (3.49), representing the lagrange function for a uniform field, has singularities, unless $\mathcal{G}=0$, $\mathcal{F}>0$, corresponding to a pure magnetic field in an appropriate coordinate system. This is the analytic expression of the fact that pairs are created by a uniform electric field. In particular, for $\mathcal{G}=0$, $-2\mathcal{F}=\mathcal{E}^2>0$, which invariantly characterizes a pure electric field, the lagrange function proper time integral,
>
> $$\mathcal{L} = \tfrac{1}{2}\mathcal{E}^2 - (1/8\pi^2)\int_0^\infty ds\, s^{-3} \exp(-m^2 s)$$
>
> $$\times [e\mathcal{E}s \cot(e\mathcal{E}s) - 1 + \tfrac{1}{3}(e\mathcal{E}s)^2], \quad (6.39)$$
>
> has singularities at
>
> $$s = s_n = n\pi/e\mathcal{E}, \quad n = 1, 2, \cdots. \quad (6.40)$$
>
> If the integration path is considered to lie above the real axis, which is an alternative version of the device embodied in Eq. (6.32), we obtain a positive imaginary contribution to $\mathcal{L}$,
>
> $$2\,\mathrm{Im}\mathcal{L} = \frac{1}{4\pi}\sum_{n=1}^\infty s_n^{-2} \exp(-m^2 s_n)$$
>
> $$= \frac{\alpha^2}{\pi^2}\mathcal{E}^2 \sum_{n=1}^\infty n^{-2} \exp\left(\frac{-n\pi m^2}{e\mathcal{E}}\right). \quad (6.41)$$
>
> This is the probability, per unit time and per unit volume, that a pair is created by the constant electric field.

Fig. 3

solved; that is, the variation can be expressed as the total variation of a quantity and therefore the variational equation integrated to get an expression for the effective action. And this, of course, is the one-loop effective action to all orders in a constant electric, or magnetic, or both, fields.

One result of this paper is the calculation of electron pair creation by a constant electric field. I read this paper when I was a student, and I really thought this was clever, amazing. I showed the relevant paragraph to Julian, and I said I thought this was really quite interesting — he just smiled. So what you do is you have a parallel plate capacitor connected to a large battery which applies a large voltage (10^{17} volts per centimeter), and then virtual positron-electron pairs will be made to tunnel out of the vacuum. This is of great theoretical interest, of course, because it is an absolutely non-perturbative result. It appears as an essential singularity as a function of the coupling constant. And here's the page that gives the calculation (Fig. 3). You see it's a typical tunneling sort of amplitude involving the exponential of one over the charge and an essential singularity appears as the electric coupling goes to zero.

There are two appendices to this paper. In one appendix (Fig. 4) the magnetic moment of the electron is re-derived (I don't know how many times he re-derived it). But it's re-derived in one page starting from first principles, starting from the basic equation in the upper left hand corner of the appendix. The magnetic moment appears in the lower right hand corner of the appendix. So here is the equation involving the modified Dirac equation when we have interactions and there's a mass operator and so here's the calculation which you can all follow in the minutes I'm talking and end up with the magnetic moment at the end. In fact, it reminds me very much of his elegant lectures in which he started at the left-hand top corner of the blackboard and ended at the right-hand bottom corner of the

APPENDIX B

An electron in interaction with its proper radiation field, and an external field, is described by the modified Dirac equation,[10]

$$\gamma_\mu(-i\partial_\mu - eA_\mu(x))\psi(x) + \int (dx') M(x, x')\psi(x') = 0. \quad (B.1)$$

To the second order in e, the mass operator, $M(x, x')$, is given by

$$M(x, x') = m_0\delta(x-x') + ie^2\gamma_\mu G(x, x')\gamma_\mu D_+(x-x'). \quad (B.2)$$

Here $G(x, x')$ is the Green's function of the Dirac equation in the external field, and $D_+(x-x')$ is a photon Green's function, expressed by

$$D_+(x-x') = (4\pi)^{-2}\int_0^\infty dt t^{-2} \exp[i\frac{1}{2}(x-x')^2/t]. \quad (B.3)$$

We shall suppose the external field to be weak and uniform. Under these conditions, the transformation function $(x(s)|x(0)')$, involved in the construction of $G(x, x')$, may be approximated by

$$(x(s)|x(0)') \simeq -i(4\pi)^{-2}\Phi(x, x')s^{-2}$$
$$\times \exp[i\frac{1}{2}(x-x')^2/s] \exp(i\frac{1}{2}e\sigma F); \quad (B.4)$$

that is, terms linear in the field strengths enter only through the Dirac spin magnetic moment. The corresponding simplification of the Green's function, obtain by averaging the two equivalent forms in Eq. (3.21), is

$$G(x, x') \simeq (4\pi)^{-2}\Phi(x, x') \int_0^\infty ds s^{-2} \exp(-im^2 s)$$
$$\times \exp[i\frac{1}{2}(x-x')^2/s]\frac{1}{2}\left\{\frac{-\gamma(x-x')}{2s} + m, \exp(i\frac{1}{2}e\sigma F)\right\}. \quad (B.5)$$

The mass operator is thus approximately represented by

$$M(x, x') = m_0\delta(x-x') + [ie^2/(4\pi)^4]\Phi(x, x')\int_0^\infty ds s^{-2}\int_0^\infty dt t^{-2}$$
$$\times \exp(-im^2 s)\exp\left[i\frac{1}{2}(x-x')^2\left(\frac{1}{s}+\frac{1}{t}\right)\right]$$
$$\times \gamma_\lambda\frac{1}{2}\left\{\frac{-\gamma(x-x')}{2s} + m, \exp(i\frac{1}{2}e\sigma F)\right\}\gamma_\lambda, \quad (B.6)$$

or

$$M(x, x') = m_0\delta(x-x') + [ie^2/(4\pi)^4]\Phi(x, x')$$
$$\times \int_0^\infty ds s^{-2} \exp(-im^2 s)\int_0^s dw w^{-2} \exp[i\frac{1}{2}(x-x')^2/w]$$
$$\times [-4m - s^{-1}\gamma(x-x') + \frac{1}{2}i\{\gamma(x-x'), \frac{1}{2}e\sigma F\}], \quad (B.7)$$

in which we have replaced t by the variable w,

$$w^{-1} = s^{-1} + t^{-1}, \quad (B.8)$$

and employed properties of the Dirac matrices, notably

$$\gamma_\lambda \sigma_{\mu\nu}\gamma_\lambda = 0. \quad (B.9)$$

We shall also write

$$(x-x')_\mu \Phi(x, x') \exp[i\frac{1}{2}(x-x')^2/w]$$
$$= 2w[-i\partial_\mu - eA_\mu(x) - \frac{1}{2}eF_{\mu\nu}(x-x')_\nu]\Phi(x, x')\exp[i\frac{1}{2}(x-x')^2/w]$$
$$\simeq [2w(-i\partial_\mu - eA_\mu(x)) - 2w^2 eF_{\mu\nu}(-i\partial_\nu - eA_\nu(x))]$$
$$\times \Phi(x, x')\exp[i\frac{1}{2}(x-x')^2/w], \quad (B.10)$$

which gives

$$M(x, x') = m_0\delta(x-x') + [e^2/(4\pi)^2]\int_0^\infty ds s^{-2} \exp(-im^2 s)$$
$$\times \int_0^s dw[2m(2-w/s) + (2w/s)(\gamma(-i\partial-eA)+m)$$
$$- 2mw(1-w/s)i\frac{1}{2}e\sigma F - iw(1+w/s)$$
$$\times \{\gamma(-i\partial-eA)+m, \frac{1}{2}e\sigma F\}](x(w)|x(0)'), \quad (B.11)$$

in virtue of the relation

$$[\gamma(-i\partial-eA), \frac{1}{2}\sigma F] = 2i\gamma F(-i\partial-eA). \quad (B.12)$$

We now introduce a perturbation procedure in which the mass operator assumes the role customarily played by the energy. To evaluate $\int (dx') M(x, x')\psi(x')$, we replace $\psi(x')$ by the unperturbed wave function, a solution of the Dirac equation associated with the mass m (we need not distinguish, to this approximation, between the actual mass m and the mechanical mass m_0). The x' integration can be effected immediately,

$$\int (x(w)|x(0)')(dx')\psi(x') = \int (x|U(w)|x')(dx')\psi(x')$$
$$= \exp(im^2 w)\psi(x), \quad (B.13)$$

since $\psi(x)$ is an eigenfunction of $\mathcal{H}$, with the eigenvalue $-m^2$. Therefore, on discarding all terms containing the operator of the Dirac equation, which will not contribute to

$$\int (dx)(dx')\bar\psi(x)M(x, x')\psi(x'),$$

we obtain

$$[\gamma(-i\partial-eA) + m - \mu'\frac{1}{2}\sigma F]\psi = 0, \quad (B.14)$$

where

$$m = m_0 + (\alpha/2\pi)m\int_0^\infty ds s^{-1}\int_0^s dw s^{-1}(2-w/s)$$
$$\times \exp[-im^2(s-w)] \quad (B.15)$$

represents the mass of a free electron, and

$$\mu' = (\alpha/2\pi)emi\int_0^\infty ds\int_0^s (dw/s)(w/s)(1-w/s)$$
$$\times \exp[-im^2(s-w)] \quad (B.16)$$

describes an additional spin magnetic moment. Both integrals are conveniently evaluated by introducing

$$u = 1 - w/s, \quad (B.17)$$

and making the replacement $s \to -is$, which yields

$$m = m_0 + (\alpha/2\pi)m\int_0^\infty ds s^{-1}\int_0^1 du(1+u) \exp(-m^2 us)$$
$$= m_0 + (3\alpha/4\pi)m\left[\int_0^\infty ds s^{-1}\exp(-m^2 s) + \frac{1}{2}\right], \quad (B.18)$$

and

$$\mu' = (\alpha/2\pi)em\int_0^\infty ds\int_0^1 du u(1-u) \exp(-m^2 us)$$
$$= (\alpha/2\pi)(e/m)\int_0^1 du(1-u) = (\alpha/2\pi)(e\hbar/2mc). \quad (B.19)$$

We thus derive the spin magnetic moment of $\alpha/2\pi$ magnetons produced by second-order electromagnetic mass effects.

[10] The concepts employed here will be discussed at length in later publications.

Fig. 4

blackboard after having made several such cycles. He would then open a conveniently located door and immediately leave the room.

So this was the paper "On Gauge Invariance and Vacuum Polarization" written in 1951. After 30 years this work has a continuing vitality. Well it's forty years now but I say after 30 years because I looked at the citations it received in the 1980's so that's 30 years.

Fig. 5(a)

Fig. 5(b)

Fig. 5(c)

The citations run from A to Z. Namely, from Adler to Zumino. Here they are (Fig. 5). I need a very fine pointer. Right there is Adler, S. and over here is Zumino, B., and that's for the first five years of the 1980's and then there's the second five years of the 1980's. I understand that once upon a time there was a historian of science at Stony Brook who would meet occasionally with the physicists there, and at one of these meetings he came to tell the physicists at lunch that he had made a great discovery that Schrödinger was greatly overrated because he had done citation surveys and found that the Schrödinger equation was very seldom cited these days. So, like the Schrödinger equation, this work is more often used than cited, and I think this list of citations doesn't do it justice. I can give a random example of the lack of scholarly approach by physicists in not citing appropriate work upon which theirs' is based. The one that most easily came to mind is the following one (Fig. 6):

Symmetric Space Scalar Field Theory*

DAVID G. BOULWARE AND LOWELL S. BROWN

*Department of Physics FM-15, University of Washington,
Seattle, Washington 98195*

Received October 28, 1981

Fig. 6

The published papers, of course, are by no means the whole story. Schwinger certainly had a very great impact as a teacher. I discussed this with my colleague Marshall Baker in Seattle. Marshall was an undergraduate at Harvard and, for some reason, went to Caltech for a year before returning to do the rest of his graduate work at Harvard. He said he'd bring me something. He brought me a whole sheaf of onion skin pages which consisted of copies of lecture notes that a friend of his (Paul Fenimore Cooper, Jr.) had taken of the lectures in Advanced Quantum Mechanics that Julian gave in 1954, and this a mechanical mass m_0 in order to obtain an observed mass m.

A SINGLE PARTICLE IN AN EXTERNAL CONSTANT MAGNETIC FIELD — 1ST ORDER MAGNETIC MOMENT CORRECTION.

We now include an external field A_μ^e in the action

$$j^\mu A'_{\mu\,EFF} \to j^\mu A'_{\mu\,EFF} + j^\mu A_\mu^e$$

AND NEGLECTING $A'_{\mu\,EFF}$ WE FIND OUR PREVIOUS WORK IS ALTERED BY REPLACING p_μ BY

$$p_\mu - eA_\mu = \pi_\mu$$

AND G_+^0 BY G_+

$$G_+ = \frac{1}{\gamma(p - eA) + m - i\epsilon}$$

WHERE WE USE AN OPERATOR A IN G_+. HENCE THE EQ IN THE MIDDLE OF p 239 BECOMES

$$\left[\gamma\pi + m_0 + ie^2\int\frac{(dk)}{(2\pi)^4}\gamma^\mu\frac{1}{\gamma(\pi-k)+m}\gamma_\mu\frac{1}{k^2}\right]\psi(x) = 0$$

Now
$$[\pi_\mu, \pi_\nu] = -[p_\mu, eA_\nu] - [eA_\mu, p_\nu]$$
$$= ie\,\partial_\mu A_\nu - ie\,\partial_\nu A_\mu$$
$$= ie\,F_{\mu\nu} = \text{CONST}$$

$$(\gamma\pi)^2 = \gamma^\mu \pi_\mu \gamma^\nu \pi_\nu = \tfrac{1}{2}\gamma^\mu\gamma^\nu\left(\{\pi_\mu, \pi_\nu\} + [\pi_\mu, \pi_\nu]\right)$$
$$= \tfrac{1}{4}\{\gamma^\mu, \gamma^\nu\}\{\pi_\mu, \pi_\nu\} + \tfrac{1}{4}[\gamma^\mu, \gamma^\nu]ie F_{\mu\nu}$$
$$= -\pi^2 - \tfrac{e}{2}\sigma^{\mu\nu}F_{\mu\nu}$$

Fig. 7(a)

So
$$\int \frac{(dk)}{(2\pi)^4} \frac{1}{k^2} \gamma^\mu \frac{1}{\gamma(\Pi-k)+m} \gamma_\mu$$

$$= \int \frac{(dk)}{(2\pi)^4} \frac{1}{k^2} \gamma^\mu \frac{m-\gamma(\Pi-k)}{(\Pi-k)^2 + m^2 - \frac{e}{2}\sigma F} \gamma_\mu$$

$$= -\int \frac{(dk)}{(2\pi)^4} \int s\, ds \int_0^1 du\, \gamma^\mu e^{-is[k^2 - 2k\Pi u + (\Pi^2 + m^2 - \frac{e}{2}\sigma F)u]}$$
$$\times [m - \gamma(\Pi-k)] \gamma_\mu$$

(SEE p 239 BOTTOM) Now

$$\int \frac{(dk)}{(2\pi)^4} e^{-is[k^2 - 2k\Pi u]} = \int \frac{(dk)}{(2\pi)^4} e^{-is[(k-\Pi u)^2 - \Pi^2 u^2]}$$

BUT WE CAN NOT DISPLACE THE INTEGRATION PATH AS Π IS NOT A NUMBER. HOWEVER THERE IS NO FIRST ORDER CONTRIBUTION TO THE INTEGRAL FROM Π AS THE ONLY VECTOR TO COMBINE WITH Π IS Π SO

$$\int \frac{(dk)}{(2\pi)^4} e^{-is(k-\Pi u)^2} = \int \frac{(dk)}{(2\pi)^4} e^{-isk'^2} + O(\Pi)^2 + \ldots \simeq \frac{-i}{16\pi^2 s^2}$$

(THIS JUSTIFIES THE WORK ON p 239. I SUPPOSE TO OUR APPROXIMATION $e^{A+B} = e^A e^B$.) WITH NO CONTRIBUTION FROM THE $\gamma k'$ TERM AS IT IS ODD WE OBTAIN

$$\int \frac{(dk)}{(2\pi)^4} \frac{1}{k^2} \gamma^\mu \frac{1}{\gamma(\Pi-k)+m} \gamma_\mu$$

$$= \frac{+i}{16\pi^2} \int \frac{ds}{s} \int_0^1 du\, \gamma^\mu e^{-is[-\Pi^2 u^2 + (\Pi^2 + m^2 - \frac{e}{2}\sigma F)u]} [m - \gamma\Pi(1-u)]\gamma_\mu$$

TO LOWEST ORDER

$$[\gamma\Pi + m]\psi(x) = 0$$

So

Fig. 7(b)

$$[m - \gamma\pi(1-u)]\gamma_\mu = \gamma_\mu[m + \gamma\pi(1-u)] + 2\pi_\mu(1-u)$$
$$\longrightarrow \gamma_\mu m u + 2\pi_\mu(1-u)$$

REGARDING σF AS SMALL AND EXPANDING THE EXPONENTIAL WITH

$$\gamma^\mu \sigma_{\lambda k} \gamma_\mu = i \gamma^\mu (\gamma_\lambda \gamma_k + 2 g_{\lambda k}) \gamma_\mu$$
$$= i(-4\gamma_\lambda \gamma_k - 2\gamma_k \gamma_\lambda + 2\gamma_\lambda \gamma_k - 4 g_{\lambda k}) = 0$$

WE SEE WE CAN NEGLECT σF IN THE EXPONENTIAL FOR THE $\gamma_\mu m u$ TERM. FURTHER MOVING π_μ TO THE LEFT THROUGH THE EXPONENTIAL CREATES NO σF TERMS — THE TERMS WE ARE INTERESTED IN. WITH THE EXP. ON THE ~~BOTH~~ RIGHT WE USE

$$(m - \gamma\pi)(m + \gamma\pi)\psi = (m^2 + \pi^2 - \tfrac{e}{2}\sigma F)\psi = 0$$

AND FINALLY OBTAIN (MOVING $\gamma_\mu m u$ LEFT WITH $\gamma^\mu \gamma_\mu = -4$)

$$\int \frac{(dk)}{(2\pi)^4} \frac{1}{k^2} \gamma^\mu \frac{1}{\gamma(\pi - k) + m} \gamma_\mu$$
$$= \frac{+i}{16\pi^2} \int_{-1}^{\,} s\, ds \int_0^1 du \left\{ -4 m u\, e^{-is[-\pi^2 u^2 + (\pi^2 + m^2)u]} + 2\gamma\pi(1-u) e^{-is[-\pi^2 u^2]} \right\}$$
$$= \frac{+i}{16\pi^2} \int \frac{ds}{s} \int_0^1 du \left\{ -4 m u\, e^{-is[m^2 u^2 + \tfrac{e}{2}\sigma F u(1-u)]} - 2m(1-u) e^{-is[m^2 u^2 - \tfrac{e}{2}\sigma F u^2]} \right\}$$

USING THE APPROXIMATE 1ST AND 2ND ORDER DIRAC EQ'S FOR THE SECOND FORM.

Fig. 8(a)

EXPANDING TO FIRST ORDER IN σF WE OBTAIN

$$\{\quad\} = -e^{-ism^2u^2}\left[+4mu + 2m(1-u)\right.$$
$$\left. -is \cdot \frac{e}{2}\sigma F\left(4mu^2(1-u) \mp 2mu^2(1-u)\right)\right]$$
$$= -e^{-ism^2u^2}\left[2m(1+u)\right.$$
$$\left. -is\, 2mu^2(1-u)\frac{e}{2}\sigma F\right]$$

AND USING THE RESULTS ON pp 240-1 FOR THE FIRST TERM WE HAVE

$$\int \frac{(dk)}{(2\pi)^4}\frac{1}{k^2}\gamma^\mu \frac{1}{\gamma(\pi-k)+m}\gamma_\mu = -\frac{im}{8\pi^2}3\ln\frac{M}{m}$$
$$-\frac{m}{8\pi^2}\frac{e}{2}\sigma F \int_0^\infty ds \int_0^1 du\, u^2(1-u)e^{-ism^2u^2}$$

$$= -\frac{im}{8\pi^2}3\ln\frac{M}{m} - \frac{m}{8\pi^2}\frac{e}{2}\sigma F \int_0^1 du \frac{(1-u)}{im^2}$$
$$= -\frac{im}{8\pi^2}3\ln\frac{M}{m_0} - \frac{1}{i}\frac{1}{(4\pi)^2}\frac{e}{2m_0}\sigma F$$

AND PUTTING THIS RESULT IN THE EQ FOR $\psi(x)$ ON p 242 WE FIND

$$\left[\gamma\pi + m - \frac{\alpha}{2\pi}\frac{e}{2m}\frac{1}{2}\sigma_{\mu\nu}F^{\mu\nu}\right]\psi(x) = 0$$

THUS THE EFFECT OF THE CURRENT SELF INTERACTION ON A SINGLE PARTICLE MOVING IN A CONSTANT EXTERNAL FIELD (BESIDES THE FIELD'S MASS CHANGE) IS TO ADD AN INTERACTION

$$-\frac{\alpha}{2\pi}\frac{e}{2m}\frac{1}{2}\sigma_{\mu\nu}F^{\mu\nu}$$

FOR A CONSTANT MAGNETIC FIELD THIS TERM IS

Fig. 8(b)

friend of Marshall's had taken notes, making a copy on this onion skin paper underneath. He would bundle them up once a week and send them to Marshall who would study them diligently in exile at Caltech.

I happen to have some student notes from a 1957 course that Schwinger gave on Quantum Field Theory and there again the magnetic moment of the electron is derived in four pages of notes. I guess that's about one *Physical Review* page. I don't remember the particular lecture. I suspect that this took something like an hour which would be two-thirds of a lecture. So here are the four pages of the student notes (Figs. 7 and 8). The notes start with essentially the same equation that started the appendix that I showed you before. Then there are manipulations to put integrals in parametric form. We have again a proper time representation to exponentiate the denominator. Then, magically looking at the right terms to look at, you get the right answer after doing the gamma matrix algebra, expanding things, and doing the integral and ... well, there's the answer. There's the anomalous magnetic moment of the electron.

At one time Schwinger was an advocate of field theory in Euclidean space time. He presented his work at the 1958 Annual International Conference on High Energy Physics which was held at CERN, the Rochester meeting at CERN that year. (The meetings were annual back then, not biannual as now.) This lecture on "Four Dimensional Euclidean Formulation of Quantum Field Theory" started out:

> I want to discuss briefly a possible avenue for the future development of quantum field theory, which I believe may be fruitful. We are all accustomed to the idealization that accompanies the quantum theory of fields in its representation of physical phenomena, i.e., the characteristic quantum mechanical feature of the use of abstract vectors and operators that symbolize physical quantities. But in one respect, at least, the quantum field theory has been conservative. It

continues to make use of a classical space-time background upon which the quantum description is superimposed. I would like to suggest a slight deepening of the abstract basis for the representation of the physical phenomena which is a replacement of the Lorentz or Minkowski space by Euclidean space.

And then goes on the body of the lecture and formula after formula. The lecture concludes:

Although we have emphasized the fundamental implications of the Euclidean representation, it will be evident that the Euclidean-type Green's functions also have practical advantages. Indeed, the utility of introducing a Euclidean metric has frequently been noted in connection with various specific problems, but an appreciation of the complete generality of the procedure has been lacking.

The lecture elicited a comment from the chairman of the session. After the lecture that chairman said:

I thank you very much for this inspiring report. To open the discussion I wish to say that for the audience it is perhaps a bit more interesting than for the speaker that the idea of analytical continuation has been anticipated by Wightman. Instead of more general transformations the speaker has selected a particular case, of rotation of 90 degrees, and I hope I interpret him correctly that he means that this has a special significance for physics and for the formalism in that particular case.

That chairman was Wolfgang Pauli.

Well, the rich vacuum structure of quantum field theory that we now understand arises from instantons which exist in Euclidean space time, and the thermodynamics of quantum field theory is really Euclidean theory, and so forth, and it's Euclidean this and Euclidean that, and I think that history has certainly proven that Schwinger was right and Pauli wrong.

Schwinger was not interested in formalism for formalism's sake. He certainly had a marvelous appreciation and he expressed things in a marvelously beautiful fashion and certainly put results in elegant forms, but he didn't do that just for its own sake. He certainly did compute numbers. Of course, the most famous example is the magnetic moment of the electron where the first five digits are the digits due to Schwinger and the later digits are due to heroic later work.

$$(g-2)/2 = 0.001\ 159\ 652\ 140\,.$$

But there are many other numbers as well. For example, there's a 1951 paper, "On a Phenomenological Neutron-Proton Interaction" by Herman Feshbach and Julian Schwinger. After a long table of numbers appears the statement:

> These calculations were performed on the Mark I calculator of the Harvard Computation Laboratory.

I'd like to conclude this talk with a story, a story of a wrong number that has to do with Schwinger's break from Cadillacs when he had an Iso Rivolto. You'll all recall this car which appeared in *Road* and *Track* in 1964. So here's the car (Fig. 9). See what it says: "It's the car we've all been looking for: Fast, comfortable, handsome, room for four $\cdots$ and now (pity) we can't afford it!" I wasn't quite sure about the exact model, so I had an email communication with Gerry Guralnick who replied to my query as follows:

> It was an Iso Revolto. It had a small block Corvette V-8 in it.

An Important Schwinger Legacy: Theoretical Tools 151

Fig. 9

I had a wonderful experience with him when he first picked it up. Heisenberg had organized a small conference in July 1965 in Feldafing (near Munich) at a resort hotel. It was what I expected of a German resort at that time — it had a fully equipped operating room.

The conference itself was quite something with an interesting mix of people, including Edward Teller. Ken Wilson presented what was to become his prize winning work. I presented what was to become known as the Higgs phenomena.

We were both beat up a bit, although I got it far worse than Ken — or at least so it seemed.

In any event, Julian grabbed me enthusiastically in the lobby (not his usual reticent self) and told me about his wonderful new car and invited me to go for a ride with him right then. As we were walking out of the hotel, Mrs. Teller came up and started asking him about his new car, and he asked her to come along. This put me in the tiny back seat. Julian drove the thing well and with enthusiasm, clearly running in Demo mode. It was actually rather impressive. After a bit Mrs. Teller piped up and said, "Julian, in the United States when we buy such an expensive car we at least expect to get an automatic transmission." She definitely scored. I will remember the look on Julian's face for the rest of my life. As I recall, his response was silence.

The following numbers are not wrong (Fig. 10). Here are the specifications for the car. It could go from 0 to 100 in 21.1 seconds. You can see the graph of speed versus elapsed time.

The wrong number was the car's license plate:

$$137\,039.$$

This I discovered when the car was in a parking lot outside of the Eastern Theoretical Physics Conference. I believe it was in 1966 and I think it was at Brown University. Since I was a rather brash kid at the time, I got a piece of paper, pasted it on the license plate and corrected the number:

$$137\,036.$$

At any rate, just remember that if you encounter technical difficulties you can really be rescued by using Schwinger techniques (Fig. 11).

An Important Schwinger Legacy: Theoretical Tools 153

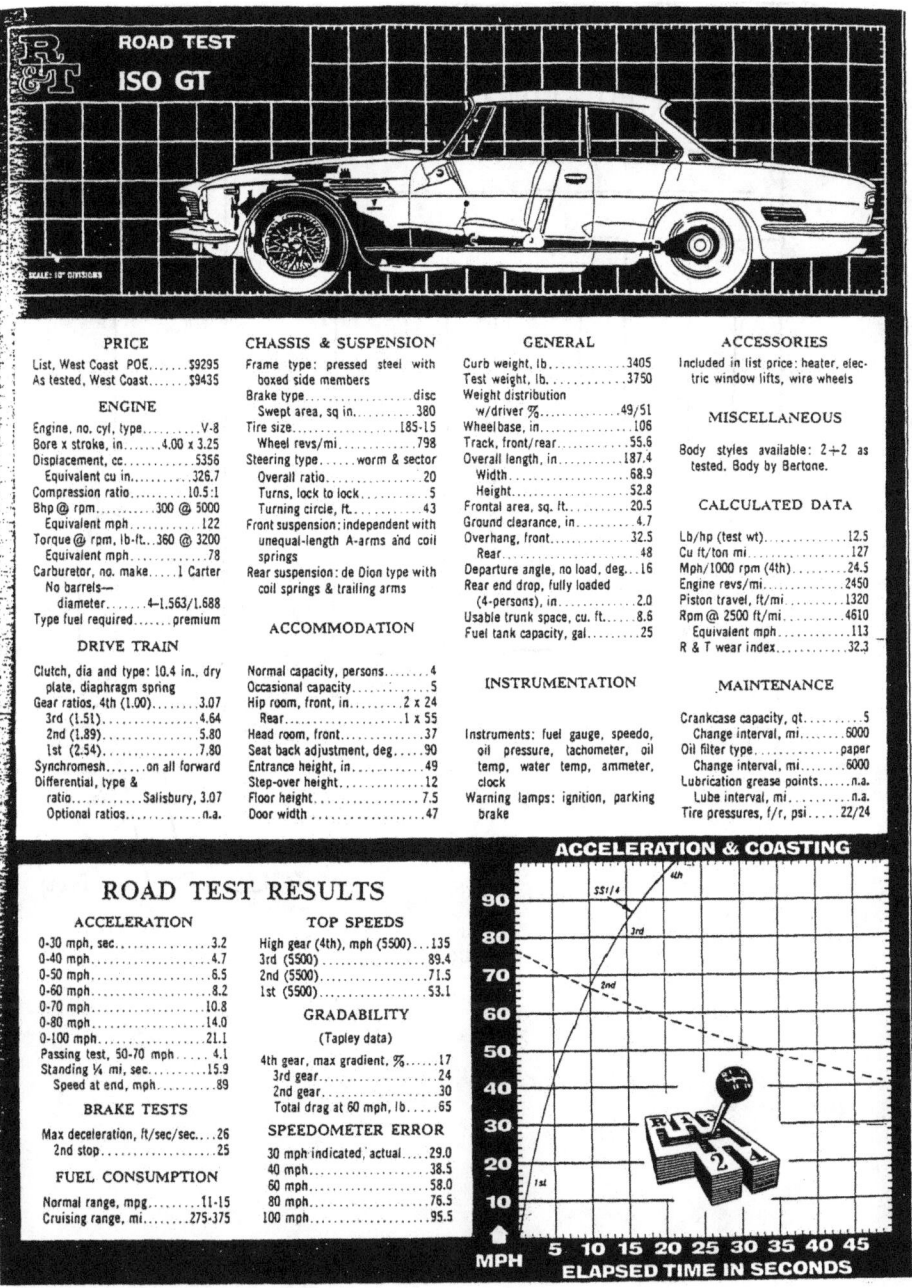

Fig. 10

Fig. 11

THE ROAD TO ELECTRO-WEAK UNIFICATION[*]

Sheldon L. Glashow

Harvard University

This is a very personal account of my interactions with Julian during the period 1954–1958, whilst I was a graduate student at Harvard. Julian was my thesis advisor, my mentor, and my professional idol. He continued to be my dear friend until his passing.

Let me begin with a joke that many of you have heard before, but which now seems appropriate. Once upon a time, a fox came upon a rabbit who was typing away in the middle of the forest. "What do you think you are doing?" asked the fox. "I am writing my thesis on how rabbits eat foxes," replied the rabbit. "Nonsense!" said the fox, "Rabbits don't eat foxes; foxes eat rabbits!" "Just take a peek in my cave," challenged the rabbit. The fox entered the rabbit's cave and was never seen again. Some time later a wolf came to the rabbit, who was still writing his thesis. "What do you think you are doing?" said the wolf, and a similar interchange ensued. The wolf entered the cave and was never seen again. Finally a ferocious grizzly bear came to chat with the rabbit. "I am writing my thesis on how rabbits eat bears," said the rabbit; "Nonsense!" growled the great beast, who was sent to the cave never to be seen again. A wise owl

---

[*]Talk in the Schwinger Memorial Session at the Washington APS-AAPT meeting on April 20, 1995

watched these strange goings-on and was puzzled. Softly sneaking into the rabbit's cave, he came upon a neat pile of fox bones. A bit further on, he discovered a neat pile of wolf bones. Finally, at the back of the cave behind a neat pile of bear bones, sat a very fat and satisfied lion picking his teeth with a bear claw. The moral of my story is that it really doesn't matter what your thesis subject is. What counts is your choice of an advisor.

Having been rejected by Harvard College, I attended Cornell, along with my high-school chum Steve Weinberg. Steve went on to graduate study at Princeton. My own Princeton interview with Wightman and Wigner went poorly. They informed me that I did seem to know the language of physics, but I did not understand its substance. Fortunately, I was admitted by Harvard this time around, where Julian was beginning to address the fascinating phenomena being observed at high energy. But I had to take all these other courses. For example: From Walter Selove, I learned that nuclear physics was not my cup of tea. Norman Ramsey taught us how atoms do complicated things in complicated electric and magnetic fields, but I couldn't care less. Paul Martin began his teaching career with an incomprehensible course on general relativity: it was clear that he would someday make a fine dean. Mercifully, I could audit courses on Japanese music and comparative literature that made up for the lapses in Cornell's liberal arts program (and my collegiate infatuation with math, physics, poker, and billiards).

So much for the dross: Schwinger was the central feature of my Harvard years. In every one of my eight semesters, I sat in the front row and listened raptly to Julian's virtuoso performances. And I learned the substance of physics. After a year or two of study, I felt prepared to ask to become his research student. Along with a dozen of my similarly enchanted peers, I dared venture into Schwinger's office to put the question. Somewhat put off by an invasion of the masses, Schwinger assigned us all a test problem to work out at home. (I think it was to express the photon propagator in Coulomb gauge.) No doubt, he hoped that we might return in a thinner and more manageable stream. A subset of us attacked the problem immediately and collaboratively and returned to the master's office at the very next opportunity: "Can we become your students?" we said. And so we did.

There we were, nervously standing in a line along the blackboard to await our calling. Danny Kleitman was first and received a personalized problem pertaining to the strong force: Julian had recently proposed a model of strong interactions called "global symmetry" and based on the group $SU(2) \times SU(2)$. Danny was to search for a mass formula linking the masses of hyperons and nucleons. Indeed, he would find such a formula, but unlike that of Gell-Mann and Okubo, it didn't work very well. Julian had exactly the right idea, but the wrong group: flavor $SU(3)$ would emerge just a few years later. Danny ultimately graduated, spent a post-doctoral year with me in Copenhagen, and became a well-known M.I.T. mathematician and my brother-in-law.

Charles Sommerfield was next in line. Charlie's problem had to do with another force of nature: electromagnetism. It was suggested that he recalculate the fourth order contributions to the magnetic moment of the electron. Julian suspected — correctly — that one of his earlier students had got it wrong. After a year or so of hard work, Charlie became the first person to get the right answer. Eventually, it would help him win a professorship at Yale.

On my turn, Julian turned to the then-mysterious weak force. Julian was convinced of the existence of an "intermediate vector boson" and of a fundamental connection between weak interactions and electromagnetism. How else to explain their common vectorial nature and their universality? My task was not precisely delineated. It was to seek and perhaps to find such a relation, and to explore its observable consequences. I remember little more of this encounter with Julian, except that it had set me on a long and treacherous voyage.

Why treacherous? In those days of yore, our understanding of the microworld was expanding at break-neck speed. A once theoretically "dictated" and experimentally "established" parity-conserving S, T, P model of the weak force was bit by bit giving way to the correct parity-violating $V - A$ picture. Schwinger's first stab at electroweak synthesis took place during a short-lived V, T interregnum. Nonetheless, he convinced himself (and me!) that a triplet of vector bosons, linked to each other as a Yang-Mills gauge theory, could possibly offer a plausible, elegant, and unified explanation of all electromagnetic and weak phenomena. Only a few vexing details remained, such as the large mass of the charged intermediary (which Steve would later provide) and its failure to conserve parity and strangeness.

Universality was the key. Schwinger interpreted universality in terms of the commutation relations of group generators; that is, of what would later become known as the algebra of charges. In short, Julian taught us the virtues of commuting together small matrices, and of relating this endeavor with the fundamental symmetries of strong, weak, and electromagnetic interactions. In the remainder of this talk, let me explain how most of my accomplishments in physics (and much of many others) grew from seeds Julian planted in his lectures, or during our many conversations in his office or over lunches at Chez Dreyfus (where he inevitably ordered steak).

Long before Lederman, Schwartz and Steinberger discovered the muon neutrino, students at Harvard knew there had to be two. Schwinger's logic was impeccable. If we choose to believe in a conserved lepton number, it would be foolish (of us or Nature) to assign it in such a way that negative electrons and muons were not distinguished. Thus, if electrons are leptons, so are *positive* muons. And as day follows night, the electron neutrino cannot be the same as the muon neutrino. Julian concluded that the electron, the neutrino, and the positive muon form a weak isospin triplet. Once again, the basic idea was right, but the details were not. There is such a thing as weak isospin, but the leptons form doublets (three of them!), not triplets.

Late in the Spring of 1958, I had assembled what I hoped might pass for a thesis. By that time, Schwinger had decamped for a summer in Madison, Wisconsin, where my thesis defense was to be held. The examining committee consisted of Bob Sachs, Paul Martin, Frank Yang, and Julian. During my presentation, Yang asked me what it could mean to say that electron and muon neutrinos were not the same. At that point, Julian took over the discussion and my safe passage was ensured. A few days later, after a somewhat raucous celebration, while Julian and I were sitting in his (pre-Lancia, pre-Iso) baby-blue Cadillac on a quiet street in suburban Madison talking physics, we were water-bombed by some irate citizens.

Having earned my degree, I planned to spend a year in the Soviet Union. I set out for the Bohr Institute in Copenhagen, where I would await the promised visa which would never come. So I spent my time recommuting my small matrices, just as Julian had taught me. The trouble was that the algebra of charges couldn't deal with parity-conserving electromagnetism and parity-violating weak interactions — unless (and it took me over a year to see this, since I no longer had direct access to Julian) the group was made just a wee bit bigger. It was only a small step from Julian's lepton triplet to two lepton doublets, from his $SU(2)$ model to Nature's $SU(2) \times U(1)$ theory, and thence, to the Z^0 boson. The imprinting had been done at Harvard.

Shortly afterward, Cabibbo invented his angle. Strangeness conservation goes with the cosine, violation with the sine. The sum of their squares is one, but why is it the same one as seen in leptonic processes? From Schwinger's viewpoint, the answer is obvious, but Gell-Mann first pointed it out, crediting me. In 1964, after a brief visit to Harvard, I returned to Copenhagen, where James Bjorken and I worried about the Cabibbo matrix, which is sort of a 3×3 matrix pretending to be 2×2. It would be much prettier, we thought, as a 4×4 matrix. Then there could be a precise analogy between leptons and quarks. And so it was that playing around with small matrices (again!) led us to charm; and somewhat later, to the explanation of the absence of strangeness-changing neutral currents. And so it has been ever since, up to and including the 5×5 matrices of grand unification. Throughout the four decades since my first meeting with Julian, whenever I accomplish something that turns out to be right, I sense that Julian is complimenting me, and at the same time, reminding me that he had said much the same thing decades ago ... and I think he is right.

JULIAN SCHWINGER: SOURCE THEORY AND THE UCLA YEARS

From Magnetic Charge to the Casimir Effect[*]

Kimball A. Milton

Imperial College and the University of Oklahoma

Abstract

Julian Schwinger began the construction of Source Theory in 1966 in response to the then apparent failure of quantum field theory to describe strong interactions, the physical remoteness of renormalization, and the utility of effective actions in describing chiral dynamics. This development did not meet with wide acceptance, and in part for this reason Julian left Harvard for UCLA in 1971. This nonacceptance was quite understandable, given the revolution in gauge theories that was then unfolding, a revolution, of course, for which he had laid much of the groundwork. Acceptance of his ideas was further impeded by his rejection of the quark model of hadrons and of QCD. I will argue, however, that the source theory development was not really so abrupt a break with the past as Julian may have implied, for the ideas and techniques in large measure were present in his work at least as early as 1951. Those techniques and ideas are still of fundamental importance to

---

[*]Invited Talk at Joint APS/AAPT meeting, Washington, April 1995

theoretical physics, so much so that the designation "source theory" has become superfluous. Julian did a great deal of innovative physics during the last 30 years of his life, and I will touch on some of the major themes, including magnetic charge, chiral dynamics, radiation theory, Thomas–Fermi models, theory of measurement, and the Casimir effect, as well as various forays into phenomenology. The impact of much of this work is not yet apparent.

Julian Schwinger, who died rather suddenly last July, was arguably the most significant theoretical physicist since Dirac. It is with great sadness that I will attempt to summarize the second half of his career, from 1966 until his death in 1994, a time during which he published nearly 100 papers on a great range of topics. His passing was especially difficult for me because at the time I, ignorant of his illness, was on my way to work with him on sonoluminescence; instead, when I arrived at UCLA I spoke at the private memorial his wife Clarice held at their home, and updated his complete publication list, a task I had commenced seventeen years previously in connection with his 60th birthday festivities.[1]

Birth of Source Theory

I begin the tale with "magnetic charge" because his last "operator" field theory papers [129,130,133,134], published in 1966, and the first "new" papers after source theory was established [147,150], published in 1968–69, were devoted to that subject. But probably the appropriate starting point is, in fact, his Nobel lecture, delivered on December 11, 1965 [132]. He ends the lecture with a discussion of phenomenological relativistic quantum field theory, and

---

[1]M. Flato, C. Fronsdal, and K. A. Milton, *Selected Papers (1937–1976) of Julian Schwinger* (Reidel, Dordrecht, 1979). The numbers in this article enclosed in square brackets refer to the list of Julian's papers I had compiled in that reference, and updated in August of 1994. The updated list is attached (see Appendix 2 — editor's note).

states that "One has still to appreciate the precise rules of phenomenological relativistic field theory,..., given that the strong fundamental interactions have operated to compose the various physical particles." Is this not a prefigurement of his attempt to create a source-theory revolution six months later?

It surely was the difficulty of incorporating strong interactions into field theory that led to "Particles and Sources," received by the *Physical Review* in July 1966 [135], a recording of lectures Julian gave in Tokyo that summer. Particle phenomenology is primary, and I personally note with relish that he cites my Oklahoma colleague George Kalbfleisch in the second sentence of the introduction for the discovery of the η' meson. This paper already included particles of all spins through the use of multispinors. The following year there was an explosion of partial (chiral) symmetry papers [137–41,143–45]. I believe that it was, in fact, his attempt to put current algebra in effective Lagrangian language, together with Weinberg, which was the immediate impetus to the source-theory development. These papers were quite important at the time.

What is Source Theory?

Although Julian had invented the notion of a source at least as early as 1951, it was only in 1966 that he realized that he could base the whole machinery of particle physics on the abstraction of particle-creation and annihilation acts. One can define a free action, say for a photon, in terms of propagation of virtual photons between photon sources, conserved in order to remove the scalar degree of freedom. But a virtual photon can in turn act as a pair of electron-positron sources, through a "primitive interaction" between electrons and photons, essentially embodied in the conserved Dirac current. So this multiparticle exchange gives rise to quantum corrections to the photon propagator, to vacuum polarization, and so on. All this without any reference to renormalization or "high-energy speculations."

In its "purest" or at least original form, such source theory ideas were used to generate perturbative amplitudes in "causal" form; that is, in which real particles were exchanged between virtual sources separated in time. From this one could deduce immediately ("space-time extrapolation") the full amplitude in spectral form, that is, in what most people would refer to as a "dispersion relation." Such a direct generation of amplitudes was extremely powerful, and often allowed a completely finite calculation to be carried out. An impressive example is our calculation of the 4th-order Compton-scattering helicity amplitudes directly in double-spectral form.[2] Noncausal methods, more reminiscent of usual Feynman diagram techniques, but significantly different in spirit, were also developed, and there we showed the power of the technique by some very simple pen-and-ink calculations of 6th-order processes contributing to the electron's magnetic moment.[3]

So what is the legacy of the source-theory experience? I think it is more evolutionary than revolutionary. New techniques were introduced by Julian, principally in the causal formulation, that supplement those introduced in earlier decades, such as the proper-time technique (which everyone uses nowadays), the quantum action principle (particularly beloved by atomic physicists now), and so on, as detailed by Lowell Brown. One commonality is the emphasis of the power of differential, rather than integral techniques ("It continues to surprise me that so many people seem to accept this formal statement [the solution of the quantum action principle as a path integral] as a satisfactory *starting* point of a theory" [160]). Certainly in my own work that has been a continuing theme, even if the word source theory now occurs but rarely. I interpret the decision of the PACS indexers to remove the "source theory" category not as a sign that source theory has become irrelevant or redundant (in the British sense); but rather

---

[2]K. A. Milton, L. L. DeRaad, Jr., and W.-Y. Tsai, *Phys. Rev.* **D6**, 1411 (1972).
[3]K. A. Milton, L. L. DeRaad, Jr., and W.-Y. Tsai, *Phys. Rev.* **D9**, 1809, 1814 (1974).

that these useful techniques are part of the common language and ammunition that theorists use to attack the most difficult problems in physics.

Let us return to the history.

Source Theory at Harvard

In 1967 "Source and Electrodynamics" [142] was published, which put QED into the new framework. The following year, Julian treated gravitons, and he gave *his* demonstration that full general relativity is essentially a consequence of assuming that the mediator of the gravitational force is a massless helicity-2 particle [146,162,163,177]. It was roughly at this point that I entered the picture, when, as a second-year student, all fear and trembling, I asked Julian if I could work for him. (But I was well prepared, bringing a good knowledge of Green's functions from the University of Washington.) I told him I was also taking Sydney Coleman's field theory lectures and Arthur Jaffe's constructive field theory course, but that was all right with Julian, in spite of his plea for the mind not "warped ... past the elastic limit." (The quotation is from the preface of [153].) The first book treatment of source theory, based on the Brandeis lectures, appeared in 1969 [149]; Julian presented me with a copy for successfully passing my oral exam (which I recall as primarily an argument between Julian and Paul Martin). I also recall the excitement of his source theory treatment of magnetic charge [147], particularly his speculative dyon model of matter which he published in Science in 1969 [150]. (His philosophy here was summed up in his quotation from Faraday: "Nothing is too wonderful to be true, if it be consistent with the laws of nature, and in such things as these, experiment is the best test of such consistency," which I would later find emblazoned on the walls of the old physics building at UCLA, Kinsey Hall.)

Three other books came out in as many years: *Discontinuity in Waveguides* (1968) [148], based on Dave Saxon's notes recording a small portion of his wartime radar work; *Quantum Kinematics and Dynamics* (1970) [152], an unfinished textbook on quantum mechanics, and *Particles, Sources, and Fields, Vol. 1* (1970) [153]. The latter was intended to be a comprehensive treatment of source theory, based on the motto "if you can't join 'em, beat 'em." Harold, the "hypothetical alert reader of limitless dedication," makes his appearance, and unlike a real student, is allowed to interrupt, particularly when he has "an historical gleam in his eye." Julian started writing the second volume of this book during a six-month sabbatical in Tokyo in 1970; on his return, he announced to his twelve or so graduate students that he was leaving Harvard in February 1971 for UCLA. Although I had only begun my fourth year at Harvard, I didn't have long to worry, for half an hour later he informed me, Lester DeRaad, Jr., and Wu-Yang Tsai that he had arranged with UCLA to bring us along as postdocs. Little did I guess that my affiliation with UCLA would last a decade!

Source Theory at UCLA

Why did he leave Harvard in 1971? Certainly, he perceived a chilly reception for source theory at Harvard, and thought (more or less erroneously) that UCLA would be more hospitable. But, probably at least as important was the fact he had been at Harvard for 25 years, and felt the need of a change. The sunny climes of Southern California, where he could and did swim and play tennis every day were an enormous attraction. Although it was billed as a temporary move, it was always clear to me that it was to be a permanent change. Appropriately, LA greeted his arrival with a major earthquake. He soon bought a beautiful home in Bel Air, with magnificent views of the city and the ocean.[4] One thing Julian did not anticipate: the caliber

---

[4] He also took the opportunity to correct the error in his license plates discussed by Lowell Brown. Since California required at least one letter in vanity plates (unfortunately not Greek), he chose brevity and universality: *A137Z*.

of graduate students at UCLA was far inferior to what he was used to at Harvard. Consequently, after more than 70 Ph.D.'s at Harvard, I believe only three ever finished at UCLA (only a few more started). (I can only recall Luis Urrutia, Walter Wilcox, and Greg Wilensky.)

Of course, also in 1971 gauge theories took off again, which doomed general reception to source theory. Julian was very much aware of what was going on, and proposed his own $U(2)$ version of the "standard model" in 1972 [155], phenomenologically acceptable in those days. (Shelly Glashow has already reminded us of his fundamental work in making the electroweak synthesis possible.) [We self-styled "sourcerer's apprentices" contributed several papers to the development of the electroweak theory.] For the next two years he worked very hard on the second volume of *Particles, Sources, and Fields* (proofed scrupulously by us three), devoted to electrodynamics, which came out in 1973 [158]. Also in 1973 was the rebirth of strong-field electrodynamics, with the publication of "Classical Radiation of Accelerated Electrons. II. A Quantum Viewpoint" [156], the first paper in which series having been published in 1949 [56]. (This illustrates the continuity of Julian's work, a subject to which I will return.) This led to a series of papers with Tsai [159,176,186], the last of which harkens back to a 1954 paper on the quantum corrections to synchrotron radiation [78]. What Julian viewed as a prediction of J/ψ, in the form of a proposal of an alternative mechanism for avoiding strangeness-changing neutral currents, appeared in the same year [157], which, after the November revolution, was followed by a series of related phenomenological forays on the ψ particles [166,169–71,173]. In 1974 he wrote two papers on "renormalization group without renormalization group" [164–65].[5]

With some very impressive work on electrodynamics (including methods harking back to his 1951 "Gauge Invariance and Vacuum

---

[5]Sometime around this point Clarice introduced me to the lovely and talented Margarita Baños, daughter of fellow physicist and Radiation Lab colleague Alfredo Baños. We were married three years later.

Polarization" paper [64] and other classic papers, an independent calculation of the 4th-order contribution to the electron's magnetic moment, and a revisiting of the axial-vector anomaly which he had discovered in [64]) constituting the first half of the third volume of PSF, he abandoned work on the book at the point where he had to face up to strong interactions. (The uncompleted third volume eventually came out in 1989 when Addison-Wesley repackaged the whole set [211].) However, he was not about to abandon high-energy physics, for in 1974 Julian continued his iconoclastic interpretation of phenomenology with an alternative viewpoint of deep-inelastic scattering based on double spectral forms (the precursor was the Deser-Gilbert-Sudarshan representation[6]), work which continued until 1977 [167,178,179,179a,181–83], starting from the valid premise that scaling does not necessitate point-like constituents.

Julian revisited magnetic charge in 1975 [172], just in time to hope that "the Price might be right" (paraphrased from *Selected Papers*).[7] A joint analysis of "dyon-dyon scattering" followed in 1976 [180]. He also became interested in the Casimir effect in 1975 [174], I think through conversations with Seth Putterman. We wrote some joint papers on the Casimir effect in 1978, among other things reconfirming Tim Boyer's surprising result on the sign of the spherical effect [187–88]. In 1977 or '78 Julian invited Stan Deser to UCLA to give us some private lessons on supersymmetry; although he submitted a paper on the multispinor basis of supersymmetry in 1978 [190], he kicked himself for not thinking of the idea first: In his words, "All right, wise guy! Then why didn't you do it first?"

During all these years he taught brilliant graduate and undergraduate courses in field theory (source theory) and quantum mechanics, lecturing for two hours a day, twice a week, followed by lunch with Bob Finkelstein and us. At first we ate at various Chinese

---

[6]S. Deser, W. Gilbert, and E. C. G. Sudarshan, *Phys. Rev.* **115**, 731 (1959).

[7]Julian often had the television on while doing physics.

restaurants, but then, as he became more diet conscious, at the Chatam in Westwood, where he always ordered rare roast beef. Tennis with Lester DeRaad was a regular part of his weekly regime.

I think it was in 1977 that Julian taught graduate electrodynamics, in a typically novel and very insightful way (including variational principles, of course, but especially noteworthy for the preëminence of physics over mathematics), and I suggested we turn the notes into a textbook. We completed a first draft (more properly, version 1.5) of a manuscript, all neatly typed by Gilda Reyes of UCLA, and signed a contract with W. H. Freeman. Unfortunately, about the time I left UCLA in 1979 Julian decided the manuscript did not sound enough like himself, and started rewriting, resulting in turgidity. The project was abandoned in 1981. However, I taught electrodynamics last fall (scheduled before Julian's death), and will do so again next year, so I have hope of reviving the book.

The Last 15 Years

In 1980, after teaching a quantum mechanics course (a not-unusual sequence of events), Julian began a series of papers on the Thomas–Fermi model of atoms [192–96,201–6]. He soon hired Berthold-Georg Englert replacing me as a postdoc to help with the elaborate calculations. This endeavor lasted until 1985.[8] In 1985 his popular book on relativity, *Einstein's Legacy*, appeared, based on a series of television programs he presented for the Open University in the UK some years earlier. (Another legacy of those programs was the robot who graced his living room thereafter.) He wrote

---

[8]I understand from conversations after my talk in Washington that this work not only is regarded as important in its own right by atomic physicists, but has led to some significant results in mathematics. A long series of substantial papers by C. Fefferman and L. Seco has been devoted to proving his conjecture about the Z dependence of the ground state energy of large atoms [193], starting with *Bull. Am. Math. Soc.* **23**, 525 (1990) and continuing through *Adv. Math.* **111**, 88 (1995).

three "Humpty Dumpty" measurement theory papers (dealing with spin coherence in a Stern–Gerlach interferometer) in 1988 [208–10], in collaboration with Marlan Scully and Englert. Those who have taken his quantum mechanics courses know how central the Stern–Gerlach experiment was to his formulation of quantum mechanics. He seemed to be spending a great deal of time on several book projects, but to my knowledge, nothing was completed. He also wrote three very interesting homages in the 1980's: "Two Shakers of Physics" [200], the pun in the title referring to himself and Tomonaga, "Hermann Weyl and Quantum Kinematics" [208a] in which he acknowledges his debt to one of his "gods," whose ways "are mysterious, inscrutable, and beyond the comprehension of ordinary mortals," and "A Path to Electrodynamics" [212], dedicated to Richard Feynman. In 1989 he began a series of papers on cold fusion [213–4,216–20]. His last physics endeavor, as I implied above, was the suggestion that the puzzling phenomenon of sonoluminescence may be due to the "dynamic Casimir effect" [221–8,230]. (The last paper was submitted on February 25, 1994, and published in the mouth of his death.) Typically, he was unaware of some of my own papers relevant to the subject, but, atypically, he was very explicitly seeking my collaboration in the last year of his life (I talked to him at some length in December 1993, at the annual Christmas party given by the Baños', which he and Clarice always attended, and at a subsequent lunch). He felt that "carrying out that program is — as one television advertiser puts it — job one" [229]. Jack Ng and I are indeed in the process of doing just that.

Conclusions

How do we place this portion of Julian's career in context? It seems to me that a number of general conclusions may be drawn.

1. I would argue that source theory was not so abrupt a break with the past as Julian presented it. It becomes increasingly clear as one reads PSF, or his general *oeuvre*, that he returns to techniques

he invented in the 1940's and 1950's. Examples are "non-causal methods" which can be found in his famous 1951 "Gauge Invariance and Vacuum Polarization paper" [64], strong field methods, which go back to his early work on synchrotron radiation [56,78] (and also GIVP), and even the theory of sources, which he introduced also in 1951 [66]. He, of course, was aware of this continuity; but he felt the need to emphasize a rather complete break. He saw a great improvement in conceptual clarity, for when he did operator field theory he carried around a great deal of baggage (which *really* is essential) which most people had dispensed with or ignored. Source theory enabled *Julian* to dispense with the "physical remoteness" [153] of renormalization and confront the physics directly. Undoubtedly, with hindsight, we can say that his later work would have had much greater impact if he had not drawn such an exclusive distinction.

2. Of course, probably a bigger impediment to the reception of his ideas was a change in the times. Dispersion relations had died before he mounted his attack, and field theory was reborn with the discovery by 't Hooft that gauge theories of weak and strong interactions made sense. He could accept the electroweak synthesis (to which he had contributed so much), but not quarks and QCD. The notion of "particles" which were not asymptotic states was too distasteful. (Yet his idea of dyons was not so different — maybe it was just the "unmellisonant" name [150].)

3. In many of his later projects, the first paper in the series was far and away the strongest. He had a very useful idea in the first deep inelastic scattering paper [167], but thereafter the work largely reduced to fitting data with many parameters. Although I am less familiar with that work, a similar characteristic is true of the Thomas–Fermi papers (although here it is the first two papers that stand out). And in the "dynamic Casimir effect" work there is enough in these many short papers for about one

substantial article; the essential calculations have yet to be carried out (some of Julian's approximations are, I believe, erroneous); and the relevance to sonoluminescence remains to be established.

4. The last 30 years of his life were not Julian's strongest scientifically. Certainly not for lack of ability: He remained an awesome calculator and a brilliant expositor of unconventional and clever ideas. But the times had changed, and Julian was no longer the molder of ideas for theoretical physics. He is sometimes criticized for venturing into phenomenology — but in fact his first, and quite substantial, papers were phenomenological. [The unfortunate distinction between theory and phenomenology (not one that Julian ever made) is a product of the last decade or so.] Much of his criticism of QCD is quite valid — the theory remains on very tenuous ground, and is more of a parametrization than the first-principles theory it pretends to be. GUTs and strings he found outrageous not because of their theoretical failings but because he, quite rightly, found the notion of a desert between 1 TeV and the Planck scale completely unbelievable — this was, after all, his reason for inventing source theory, to separate high-energy speculations from models of low-energy phenomena.

5. As footnote 8 illustrates, we should not underestimate the power of his work to have a continuing impact. We can confidently expect future surprises. This may be true as well of the many papers in the attached list to which I have not referred, because they do not fit into a well-defined pigeonhole. I can only urge the reader to read his papers, for riches are contained therein.

Eight months before his death, Julian made his first appearance on the Internet (and his penultimate publication in any form) with his July 1993 Nottingham lecture, "The Greening of Quantum Field Theory: George and I" (hep-ph/9310283) [229]. This lecture provides a remarkable overview of Julian's work from his own perspective. I commend his final words to you: Like George Green, "he is, in a manner of speaking, alive, well, and living among us."

Acknowledgements

I am grateful to the UK PPARC for a Senior Visiting Fellowship and Imperial College for its hospitality. I thank UCLA for its hospitality during the period when I updated Julian's publication list, and the US Department of Energy for partial financial support. I dedicate this article to Julian Schwinger, the most brilliant physicist I have known, and one of my very dearest friends, to whom I owe so much.

JULIAN SCHWINGER*

C. N. Yang

State University of New York at Stony Brook and Chinese University of Hong Kong

When I enrolled as a graduate student at the University of Chicago in 1946, Schwinger was already a legend. I heard many stories about him. Like the one about his virtuoso performance at the Radiation Laboratory in Cambridge, Massachusetts, and the one about his never coming to work until way after ordinary physicists' dinner time, and many others.

I also had the occasion to study his striking papers. I remember still vividly today reading in the library some early articles by him. The first one was his paper with E. Teller about the scattering of neutrons by ortho- and parahydrogen (*Phys. Rev.* **52**, 286 (1937)). Also I remember studying in great detail the paper he wrote with W. Rarita on neutron-proton interactions (*Phys. Rev.* **59**, 436 (1941)).

Then there occurred the great event of renormalization. Let me just remind you of the historical sequence of what happened. At the Shelter Island Conference, the striking results obtained by Lamb and Retherford, and by Foley and Kush were reported. That was in

---

*Lecture in the Schwinger Memorial Session of the APS-AAPT meeting in Washington D.C., April 20, 1995

June 1947. Very remarkably, within a few weeks, Bethe submitted a paper (*Phys. Rev.* **72**, 339 (1947)) which got most of the Lamb shift right. He got 1040 MHZ. And also very remarkably, within a few months, Schwinger published his important paper (*Phys. Rev.* **73**, 416 (1948)) in early 1948. I think the paper in fact was submitted at the end of 1947; it appeared in 1948. He gave the famous number $(\alpha/2\pi)$ for the extra magnetic moment of the electron. He reported on it at the New York APS meeting (January 29–31, 1948).

Then came the famous Pocono Conference (March 30–April 1, 1948). I did not make it to the meeting. I was just a graduate student. From Chicago, Fermi, Teller, and Wentzel went. Fermi did not usually take notes when he went to a conference. But this time, he took voluminous notes because he was aware that it was a historical event to listen to what Schwinger had to say. After they came back to Chicago, there was the question of how to digest these notes. Fermi gathered Teller and Wentzel and four graduate students, viz., Geoffrey Chew, Murph Goldberger, Marshall Rosenbluth, and myself, into his office, and we spent weeks trying to digest what Fermi had written down as what Schwinger had said. This lasted from April to May, 1948. Murph kept notes. I still have a copy of these; it totals 49 pages. After about six weeks of meeting several times a week in Fermi's office for something like two hours each session we were all very tired, and none of us felt that we had understood what Schwinger had done. We only knew that Schwinger had done something brilliant, namely, he had produced this $(\alpha/2\pi)$ and he was also already into the calculations of the Lamb shift.

At the end of our six weeks of work, somebody asked, "Wasn't it true that Feynman also talked?" All three said, "Yes, yes, Feynman did talk." "What did he say?" None of them could say. All they remembered was Feynman's strange notation: p with a slash in it.

Now let us take a look at the history of the theory of renormalization again. The first paper was Julian Schwinger's and I have already talked about that. The next one was a Japanese paper by H. Fukuda, Y. Miyamoto, and S. Tomonaga (*Prog. Theo. Phys.* **4**, 47, 121 (1949)). Then Schwinger came back and in 1949 he produced the 1051 MHZ (*Phys. Rev.* **75**, 898 (1949)) for the Lamb shift, which is the correct value from the theoretical viewpoint up to that order of magnitude. That was followed by Feynman (*Phys. Rev.* **76**, 749, 769 (1949)). So these three papers (by Tomonaga, Schwinger, and Feynman separately) were the papers that gave the right relativistic formulation and calculations of the Lamb shift. Another great event around that time was the series of two papers by Freeman Dyson (*Phys. Rev.* **75**, 486, 1736 (1949)), which really explained what all these things were about to a person like myself, who was a new postdoc very anxious to learn what was going on but had difficulties reading all these complicated papers. I should also mention that from the theoretical side, the people who contributed to renormalization theory included in addition Bethe, Kramers, Lamb, Oppenheimer, Weisskopf, and a number of others.

From the historical viewpoint, I would say that the renormalization development, both theory and experiment, was the first great excitement in post-war physics. It also signified the end of the monopoly of fundamental physics by Europe. It signified the beginning of a new era, the American era.

To use a metaphor: Renormalization was one of the great peaks of the development of fundamental physics in this century. Scaling the peak was a difficult enterprise. It required technical skill, courage, subtle judgements and great persistence. Many people had contributed to this enterprise. Many many people can climb the peak now. But, *the person who first conquered the peak was Julian Schwinger*.

I first met Julian in 1948 at a summer school at Michigan, which was a very famous set of schools starting from before the war. But I did not really get to know him. He was shy and it was difficult for a person that he had never met to discuss any physics with him. I really got to know him in 1958 when both of us were invited by Bob Sachs to visit Madison, Wisconsin. And that summer, Julian, his wife Clarice, my wife Chih Li, and I got together often. I learned to appreciate Julian as not only a great physicist but also a gentleman.

Schwinger had superb command of the English language. Even in casual conversation he spoke in complete, crafted, and polished sentences. A good example of that could be found in the Proceedings of the 1980 conference at Fermi Lab about the birth of particle physics. In that conference Julian gave a speech entitled "Renormalization theory of quantum electrodynamics: an individual view." This is an extremely interesting article. It gives the details of the step-by-step development of the theory of renormalization. It also clearly demonstrates Schwinger's superb command of the English language. I highly recommend it to you.

Perhaps the most striking sentence in this article is the following: *Like the silicon chip of more recent years, the Feynman diagram was bringing computation to the masses.*

I believe Schwinger was justifiably unhappy that the younger generations, dazzled by the brilliant performer that Feynman was, have forgotten that it was Schwinger who had first scaled the mighty peak that is known as renormalization.

Feynman and Schwinger were two great physicists of our times. Each had made many insightful contributions to physics. They were born in the same year — 1918. But personality-wise, they were as different from each other as any two individuals could be. I often thought that one could write a book with the title: *Schwinger and Feynman, A Study in Contrast:*

> Twenty percent impulsive clown, twenty percent professional nonconformist, sixty percent brilliant physicist, Feynman strived to be a great performer almost as much as to be a great physicist.

Shy, erudite, speaking and writing in crafted and polished sentences, Schwinger epitomized the cultured perfectionist and the quiet inward-looking gentleman.

I moved from the Institute for Advanced Study in 1966 to Stony Brook. And together with the administration of Stony Brook I made an effort to induce Schwinger to come to Stony Brook. At one point I wrote the following letter (dated April 18, 1968) to Julian:

> Just a note to say we hope to hear from you soon (and positively, I hope).
>
> May I also add one point which keeps coming to me: Harvard, prestigious as it is, cannot add to your lustre. It is you who brings lustre to whatever Institution that you choose to join.

Unfortunately for us, Julian went to UCLA. A number of years later, I received a letter from the then chairman of the physics department at UCLA, asking me to support a proposal that they were making to appoint Julian a University Professor. This is my reply (dated June 13, 1978):

> Professor Julian Schwinger is among the great physicists of the contemporary era. His work covers an amazingly wide range, from nuclear physics to elementary particle physics to field theory, from synchrotron radiation to group theory to microwave propagation. He has contributed to all of these fields important and essential ideas that have affected the whole field in the last thirty years.
>
> The most important work of Schwinger was his contribution to renormalization, a contribution

that stands among the greatest developments in physics in mid-twentieth century.

Professor Schwinger is an eminently successful teacher. He probably has graduated more influential theoretical Ph.D. students than any other living physicist. His lectures are polished, elegant and have a particular style of distinction characteristic of Schwinger's approach to physics.

Schwinger is a quiet person with great sensitivity and perception in many intellectual areas beyond the science of physics.

I support, without reservation, the proposal to appoint Julian Schwinger as University Professor at the University of California. My only amazement is that he was not appointed to such a position when he joined UCLA.

APPENDIX 1
LIST OF SCHWINGER'S DOCTORAL STUDENTS

(Based on the 1975 Directory of Harvard Physics Doctoral Alumni and information provided by the UCLA Physics Department)

Name	Ph.D. year	Name	Ph.D. year
Arnowitt, Richard L.	53	Deser Stanley	53
Aronson, Raphael	52	DeWitt, Bryce S.	50
Baker, Marshall	58	Eisenstein, Julian C.	48
Bakshi, Pradip M.	63	Engelsberg, Stanley	61
Barone, Stephen R.	67	Falk, David S.	59
Baym, Gordon A.	60	Feldman, David	49
Bernstein, Jeremy	55	Glashow, Sheldon L.	59
Boulware, David G.	62	Glauber, Roy J.	49
Brachman, Malcolm K.	49	Horing, Norman J.	64
Brown, Lowell S.	61	Horwitz, Lawrence P.	57
Case, Kenneth M.	48	Ivanetich, Richard J.	69
Chan, Lai H.	66	Johnson, Kenneth A.	55
Chang, Shau J.	67	Karagiannis, Evangelos	90
Clapp, Roger E.	49	Keilson, Julian	50
Clark, Donald	94	Kivelson, Margaret G.	57
Clark, Melville, Jr.	49	Klein, Abraham	50
DeHoffmann, Frederic	48	Kleitman, Daniel J.	58
DeRaad, Lester L., Jr.	71	Kohn, Walter	48

List of Schwinger's Doctoral Students

Name	Ph.D. year	Name	Ph.D. year
Lazarus, Roger B.	51	Raphael, Robert B.	55
Lepore, Joseph V.	48	Rohrlich, Fritz	49
Lieber, Michael	67	Sawyer, Raymond F.	58
Lippmann, Bernard A.	48	Sommerfield, Charles M.	57
Lynch, David D.	67	Stern, Adolph	52
Mahanthappa, Kalyana T.	61	Thorn, Robert N.	53
Malenka, Bertram J.	52	Tsai, Wu Y.	71
Martin, Paul C.	54	Urrutia, Luis F.	78
McCormick, Bruce H.	55	Warner, Charles III	58
Merzbacher, Eugen	50	Warnock, Robert L.	59
Milton, Kimball A.	71	Weitzner, Harold	58
Mottelson, Ben R.	50	Wilcox, Walter M.	81
Neuman, Maurice	49	Wilensky, Gregg D.	81
Newton, Roger G.	54	Yamauchi, Hiroshi	50
Ng, Y. Jack	74	Yan, Tung M.	68
Nieh, Hwa T.	66	Yao, Edward Y.P.	64
Peaslee, Alfred T., Jr.	55	Yildiz, Asim	73
Phares, Alain J.	73	Zemach, Charles	55
Radkowski, Alfred P.	69		

APPENDIX 2
JULIAN SCHWINGER — LIST OF PUBLICATIONS

This incorporates and updates the publication list found in *Selected Papers (1937–1976) of Julian Schwinger*

Edited by M. Flato, C. Fronsdal, and K. A. Milton (Reidel, Dordrecht, 1979)

0. On the Interaction of Several Electrons, unpublished (1934).
1. On the Polarization of Electrons by Double Scattering (with O. Halpern), *Phys. Rev.* **48**, 109 (1935).
2. On the β-Radioactivity of Neutrons (with L. Motz), *Phys. Rev.* **48**, 704 (1935).
3. On the Magnetic Scattering of Neutrons, *Phys. Rev.* **51**, 544 (1937).
4. On Non-Adiabatic Processes in Inhomogeneous Fields, *Phys. Rev.* **51**, 648 (1937).
5. The Scattering of Neutrons by Ortho- and Parahydrogen (with E. Teller), *Phys. Rev.* **51**, 775 (1937).
6. Depolarization by Neutron-Proton Scattering (with I. I. Rabi), *Phys. Rev.* **51**, 1003 (1937).
7. Neutron Energy Levels (with J. Manley and H. Goldsmith), *Phys. Rev.* **51**, 1022 (1937).
8. The Scattering of Neutrons by Ortho- and Parahydrogen (with E. Teller), *Phys. Rev.* **52**, 286 (1937).
9. On the Spin of the Neutron, *Phys. Rev.* **52**, 1250 (1937).
10. The Widths of Nuclear Energy Levels (with J. Manley and H. Goldsmith), *Phys. Rev.* **55**, 39 (1939).

11. The Neutron-Proton Scattering Cross Section (with V. Cohen and H. Goldsmith), *Phys. Rev.* **55**, 106 (1939).
12. The Resonance Absorption of Slow Neutrons in Indium (with J. Manley and H. Goldsmith), *Phys. Rev.* **55**, 107 (1939).
13. On the Neutron-Proton Interaction, *Phys. Rev.* **55**, 235 (1939).
14. The Scattering of Neutrons by Hydrogen and Deuterium Molecules (with M. Hamermesh), *Phys. Rev.* **55**, 679 (1939).
15. On Pair Emission in the Proton Bombardment of Fluorine (with J. R. Oppenheimer), *Phys. Rev.* **56**, 1066 (1939).
16. Neutron-Deuteron Scattering Cross Section (with L. Motz), *Phys. Rev.* **57**, 161 (1940).
17. The Scattering of Thermal Neutrons by Deuterons (with L. Motz), *Phys. Rev.* **58**, 26 (1940).
18. The Electromagnetic Properties of Mesotrons (with H. Corben), *Phys. Rev.* **58**, 191 (1940).
19. The Electromagnetic Properties of Mesotrons (with H. Corben), *Phys. Rev.* **58**, 953 (1940).
20. Neutron Scattering in Ortho- and Parahydrogen and the Range of Nuclear Forces, *Phys. Rev.* **58**, 1004 (1940).
21. The Photodisintegration of the Deuteron (with W. Rarita), *Phys. Rev.* **59**, 215 (1941).
22. The Photodisintegration of the Deuteron (with W. Rarita and H. Nye) *Phys. Rev.* **59**, 209 (1941).
23. On the Neutron-Proton Interaction (with W. Rarita), *Phys. Rev.* **59**, 436 (1941).
24. On the Exchange Properties of the Neutron-Proton Interaction (with W. Rarita), *Phys. Rev.* **59**, 556 (1941).
25. On a Theory of Particles with Half-Integral Spin (with W. Rarita), *Phys. Rev.* **60**, 61 (1941).
26. On the Interaction of Mesotrons and Nuclei (with J. R. Oppenheimer), *Phys. Rev.* 60, 150 (1941).
27. The Theory of Light Nuclei (with E. Gerjuoy), *Phys. Rev.* **60**, 158 (1941).
28. On the Charged Scalar Mesotron Field, *Phys. Rev.* **60**, 159 (1941).
29. The Quadrupole Moment of the Deuteron and the Range of Nuclear Forces, *Phys. Rev.* **60**, 164 (1941).
30. On Tensor Forces and the Theory of Light Nuclei (with E. Gerjuoy), *Phys. Rev.* **61**, 138 (1942).
31. On a Field Theory of Nuclear Forces, *Phys. Rev.* **61**, 387 (1942).
32. On the Magnetic Moments of H^3 and He^3 (with R. Sachs), *Phys. Rev.* **61**, 732 (1942).

33. The Scattering of Slow Neutrons by Ortho and Para Deuterium (with M. Hamermesh), *Phys. Rev.* **69**, 145 (1946).
34. Polarization of Neutrons by Resonance Scattering in Helium, *Phys. Rev.* **69**, 681 (1946).
35. Electron Orbits in the Synchrotron (with D. Saxon), *Phys. Rev.* **69**, 702 (1946).
36. The Magnetic Moments of H^3 and He^3 (with R. Sachs), *Phys. Rev.* **70**, 41 (1946).
37. Electron Radiation in High Energy Accelerators, *Phys. Rev.* **70**, 798 (1946).
38. Neutron Scattering in Ortho- and Parahydrogen (with M. Hamermesh), *Phys. Rev.* **71**, 678 (1947).
39. On the Radiation of Sound from a Unflanged Circular Pipe (with H. Levine), *Phys. Rev.* **72**, 742 (1947).
40. A Variational Principle for Scattering Problems, *Phys. Rev.* **72**, 742 (1947).
41. On the Radiation of Sound from an Unflanged Circular Pipe (with H. Levine), *Phys. Rev.* **73**, 383 (1948).
42. On the Polarization of Fast Neutrons, *Phys. Rev.* **73**, 407 (1948).
43. On Quantum-Electrodynamics and the Magnetic Moment of the Electron, *Phys. Rev.* **73**, 416 (1948).
44. A Note on Saturation in Microwave Spectroscopy (with R. Karplus), *Phys. Rev.* **73**, 1020 (1948).
45. On the Electromagnetic Shift of Energy Levels (with V. Weisskopf), *Phys. Rev.* **73**, 1272 (1948).
46. On the Theory of Diffraction by an Aperture in an Infinite Plane Screen. I (with H. Levine), *Phys. Rev.* **74**, 958 (1948).
47. An Invariant Quantum Electrodynamics, *Phys. Rev.* **74**, 1212 (1948).
48. Variational Principles for Diffraction Problems (with H. Levine), *Phys. Rev.* **74**, 1212 (1948).
49. On Tensor Forces and the Variation-Iteration Method (with H. Feshbach and J. Eisenstein), *Phys. Rev.* **74**, 1223 (1948).
50. Quantum Electrodynamics I. A Covariant Formulation, *Phys. Rev.* **74**, 1439 (1948).
51. Radiative Correction to the Klein–Nishina Formula (with D. Feldman), *Phys. Rev.* **75**, 358 (1949).
52. Quantum Electrodynamics II. Vacuum Polarization and Self Energy, *Phys. Rev.* **75**, 651 (1949).
53. On Radiative Corrections to Electron Scattering, *Phys. Rev.* **75**, 898 (1949).

54. On the Theory of Diffraction by an Aperture in an Infinite Plane Screen. II. (with H. Levine), *Phys. Rev.* **75**, 1423 (1949).
55. On the Transmission Coefficient of a Circular Aperture (with H. Levine), *Phys. Rev.* **75**, 1608 (1949).
56. On the Classical Radiation of Accelerated Electrons, *Phys. Rev.* **75**, 1912 (1949).
57. Quantum Electrodynamics III. The Electromagnetic Properties of the Electron — Radiative Corrections to Scattering, *Phys. Rev.* **76**, 790 (1949).
58. On the Charge Independence of Nuclear Forces, *Phys. Rev.* **78**, 135 (1950).
59. On the Self-Stress of the Electron (with S. Borowitz and W. Kohn) *Phys. Rev.* **78**, 345 (1950).
60. Variational Principles for Scattering Processes. I. (with B. Lippman), *Phys. Rev.* **79**, 469 (1950).
61. On the Theory of Electromagnetic Wave Diffraction by an Aperture in an Infinite Plane Conducting Screen (with H. Levine), *Comm. Pure Appl. Math. III*, **4**, 355 (1950).
62. Quantum Dynamics, *Science* **113**, 479 (1951).
63. On the Representation of the Electric and Magnetic Fields Produced by Currents and Discontinuities in Wave Guides. I. (with N. Marcuvitz), *J. Appl. Phys.* **22**, 806 (1951).
64. On Gauge Invariance and Vacuum Polarization, *Phys. Rev.* **82**, 664 (1951).
65. The Theory of Quantized Fields. I, *Phys. Rev.* **82**, 914 (1951).
66. On the Green's Functions of Quantized Fields. I, II, *Proc. Natl. Acad. Sci. U.S.A.* **37**, 452, 455 (1951).
67. On a Phenomenological Neutron-Proton Interaction (with H. Feshbach), *Phys. Rev.* **84**, 194 (1951).
68. Electrodynamic Displacement of Atomic Energy Levels (with R. Karplus and A. Klein), *Phys. Rev.* **84**, 597 (1951).
69. On Angular Momentum, 1952, later published in *Quantum Theory of Angular Momentum*, edited by L. C. Biedenharn and H. Van Dam, (Academic Press, New York, 1965), p. 229.
70. Electrodynamic Displacement of Atomic Energy Levels. II. Lamb Shift (with R. Karplus and A. Klein), *Phys. Rev.* **86**, 288 (1952).
71. Radiation Force and Torque (with H. Levine), *Phys. Rev.* **87**, 224 (1952).
72. On High Energy Nucleon Scattering and Isobars (with R. B. Raphael), *Phys. Rev.* **90**, 373 (1953).
73. The Theory of Quantized Fields. II., *Phys. Rev.* **91**, 713 (1953).

74. The Theory of Quantized Fields. III., *Phys. Rev.* **91**, 728 (1953).
75. A Note on the Quantum Dynamical Principle, *Philos. Mag.* **44**, 1171 (1953).
76. The Theory of Quantized Fields. IV., *Phys. Rev.* **92**, 1283 (1953).
77. The Theory of Quantized Fields. V., *Phys. Rev.* **93**, 615 (1954).
78. The Quantum Correction in the Radiation by Energetic Accelerated Electrons, *Proc. Natl. Acad. Sci. U.S.A.* **40**, 132 (1954).
79. Use of Rotating Coordinates in Magnetic Resonance Problems (with I. I. Rabi and N. Ramsey), *Rev. Mod. Phys.* **26**, 167 (1954).
80. The Theory of Quantized Fields. VI., *Phys. Rev.* **94**, 1362 (1954).
81. Dynamical Theory of K Mesons, *Phys. Rev.* **104**, 1164 (1956).
82. A Theory of the Fundamental Interactions, *Ann. Phys. (N.Y.)* **2**, 407 (1957).
83. *Quantum Electrodynamics*, Editor, Dover, New York, 1958.
84. Spin, Statistics and the TCP Theorem, *Proc. Natl. Acad. Sci. U.S.A.* **44**, 223 (1958).
85. Addendum to Spin, Statistics and the TCP Theorem, *Proc. Natl. Acad. Sci. U.S.A.* **44**, 617 (1958).
86. On the Euclidean Structure of Relativistic Field Theory, *Proc. Natl. Acad. Sci. U.S.A.* **44**, 956 (1958).
87. Four-Dimensional Euclidean Formulation of Quantum Field Theory, *Proceedings of the 1958 International Conference on High Energy Physics*, CERN, Geneva, 1958, p. 134.
88. Euclidean Quantum Electrodynamics, *Phys. Rev.* **115**, 721 (1959).
89. Theory of Many-Particle Systems. I. (with P. C. Martin), *Phys. Rev.* **115**, 1342 (1959).
90. Field Theory Commutators, *Phys. Rev. Lett.* **3**, 296 (1959).
91. The Algebra of Microscopic Measurement, *Proc. Natl. Acad. Sci. U.S.A.* **45**, 1542 (1959).
92. Field Theory Methods, 1959 Brandeis University Summer Institute in Theoretical Physics.
93. The Geometry of Quantum States, *Proc. Natl. Acad. Sci. U.S.A.* **46**, 257 (1960).
94. Field Theory of Unstable Particles, *Ann. Phys. (N.Y.)* **9**, 169 (1960).
95. Euclidean Gauge Transformation, *Phys. Rev.* **117**, 1407 (1960).
96. Unitary Operator Bases, *Proc. Natl. Acad. Sci. U.S.A.* **46**, 570 (1960).
97. Unitary Transformations and the Action Principle, *Proc. Natl. Acad. Sci. U.S.A.* **46**, 883 (1960).
98. The Special Canonical Group, *Proc. Natl. Acad. Sci. U.S.A.* **46**, 1401 (1960).

99. Field Theory Methods in Non-Field Theory Contexts, 1960 Brandeis University Summer Institute in Theoretical Physics, Lecture Notes, p. 223.
100. On the Bound States of a Given Potential, *Proc. Natl. Acad. Sci. U.S.A.* **47**, 122 (1961).
101. Brownian Motion of a Quantum Oscillator, *J. Math. Phys.* **2**, 407 (1961).
102. Quantum Variables and the Action Principle, *Proc. Natl. Acad. Sci. U.S.A.* **47**, 1075 (1961).
103. Spin and Statistics (with L. Brown), *Prog. Theor. Phys. (Kyoto)* **26**, 917 (1961).
104. Gauge Invariance and Mass, *Phys. Rev.* **125**, 397 (1962).
105. Non-Abelian Gauge Fields. Commutation Relations, *Phys. Rev.* **125**, 1043 (1962).
106. Exterior Algebra and the Action Principle. I., *Proc. Natl. Acad. Sci. U.S.A.* **48**, 603 (1962).
107. Non-Abelian Gauge Fields. Relativistic Invariance, *Phys. Rev.* **127**, 324 (1962).
108. Gauge Invariance and Mass. II., *Phys. Rev.* **128**, 2425 (1962).
109. Quantum Variables and Group Parameters, Il *Nuovo Cimento* **30**, 278 (1963).
110. Non-Abelian Gauge Fields. Lorentz Gauge Formulation, *Phys. Rev.* **130**, 402 (1963).
111. Commutation Relations and Conservation Laws, *Phys. Rev.* **130**, 406 (1963).
112. Energy and Momentum Density in Field Theory, *Phys. Rev.* **130**, 800 (1963).
113. Quantized Gravitational Field, *Phys. Rev.* **130**, 1253 (1963).
114. Quantized Gravitational Field. II., *Phys. Rev.* **132**, 1317 (1963).
115. Gauge Theories of Vector Particles, *Theoretical Physics* (Trieste Seminar, 1962), (IAEA, Vienna, 1963), p. 89.
116. Coulomb Green's Function, *J. Math. Phys.* **5**, 1606 (1964).
117. Non-Abelian Vector Gauge Fields and the Electromagnetic Field, *Rev. Mod. Phys.* **36**, 609 (1964).
118. Field Theory of Matter, *Phys. Rev.* **135**, B816 (1964).
119. *A Ninth Baryon, Coral Gables Conference on Symmetry Principles at High Energy*, 1964, edited by B. Kursunoglu and A. Perlmutter (Freeman, San Francisco, 1964), p. 127.
120. A Ninth Baryon?, *Phys. Rev. Lett.* **12**, 237 (1964).
121. $\Delta T = 3/2$ Nonleptonic Decay, *Phys. Rev. Lett.* **12**, 630 (1964).

122. Broken Symmetries and Weak Interactions, *Phys. Rev. Lett.* **13**, 355 (1964).
123. Broken Symmetries and Weak Interactions. II., *Phys. Rev. Lett.* **13**, 500 (1964).
124. Field Theory of Matter. II., *Phys. Rev.* **136**, B1821 (1964).
125. *Field Theory of Particles, Lectures on Particles and Field Theory* (1964 Brandeis Lectures), edited by S. Deser and K. Ford (Prentice-Hall, Englewood Cliffs, N.J., 1965), p. 145.
126. Field Theory of Matter, *Proceedings of the 12th International Conference on High Energy Physics*, Dubna, 1964 (Atomizdat, Moscow, 1966), Vol. 1, p. 771.
127. Field Theory of Matter. III. Phenomenological Field Theory, *Coral Gables Conference on Symmetry Principles at High Energy*, 1965, edited by B. Kursunoglu, A. Perlmutter, and I. Sakmar (Freeman, San Francisco, 1965), p. 372.
128. Field Theory of Matter. IV., *Phys. Rev.* **140**, B158 (1965).
129. Magnetic Charge and Quantum Field Theory, *Phys. Rev.* **144**, 1087 (1966).
130. Magnetic Charge and Quantum Field Theory, *Coral Gables Conference on Symmetry Principles at High Energy*, 1966, edited by A. Perlmutter, J. Wojtaszek, G. Sudarshan, and B. Kursunoglu (Freeman, San Francisco, 1966), p. 233.
131. Lectures on Quantum Field Theory, Coral Gables, 1966, University of Miami, 1967.
132. Relativistic Quantum Field Theory, Nobel Lecture, in *Nobel Lectures — Physics, 1963–1970* (Elsevier, Amsterdam, 1972).
133. Electric and Magnetic-Charge Renormalization. I., *Phys. Rev.* **151**, 1048 (1966).
134. Electric and Magnetic-Charge Renormalization. II., *Phys. Rev.* **151**, 1055 (1966).
135. Particles and Sources, *Phys. Rev.* **152**, 1219 (1966).
136. Sourcery, *Coral Gables Conference on Symmetry Principles at High Energy*, 1967, edited by A. Perlmutter and B. Kursunoglu (Freeman, San Francisco, 1967), p. 180.
137. Chiral Dynamics, *Phys. Lett.* **24B**, 473 (1967).
138. Mass Empirics, *Phys. Rev. Lett.* **18**, 797 (1967).
139. Partial Symmetry, *Phys. Rev. Lett.* **18**, 923 (1967).
140. Photons, Mesons and Form Factors, *Phys. Rev. Lett.* **19**, 115 (1967).
141. Radiative Corrections in β Decay, *Phys. Rev. Lett.* **19**, 1501 (1967).
142. Sources and Electrodynamics, *Phys. Rev.* **158**, 1391 (1967).
143. Boson Mass Empirics, *Phys. Rev. Lett.* **20**, 516 (1968).

144. Gauge Fields, Sources and Electromagnetic Masses, *Phys. Rev.* **165**, 1714 (1968); *Phys. Rev.* **167**, 1546 (1968).
145. Chiral Transformations, *Phys. Rev.* **167**, 1432 (1968).
146. Sources and Gravitons, *Phys. Rev.* **173**, 1264 (1968).
147. Sources and Magnetic Charge, *Phys. Rev.* **173**, 1536 (1968).
148. *Discontinuities in Wave Guides* (with D. Saxon), Gordon and Breach, New York, 1968.
149. *Particles and Sources*, Gordon and Breach, New York, 1969.
150. A Magnetic Model of Matter, *Science*, **165**, 757 (1969).
151. Theory of Sources, *Contemporary Physics* (Trieste Symposium 1968), (IAEA, Vienna, 1969), Vol. 11, p. 59.
152. *Quantum Kinematics and Dynamics*, Benjamin, New York, 1970.
153. *Particles, Sources and Fields*, Vol. I, Addison-Wesley, Reading, Mass., 1970.
154. Unit-Spin Propagation Functions and Form Factors, *Phys. Rev.* **D3**, 1967 (1971).
155. How Massive is the W Particle?, *Phys. Rev.* **D7**, 908 (1973).
156. Classical Radiation of Accelerated Electrons II. A Quantum Viewpoint, *Phys. Rev.* **D7**, 1696 (1973).
157. How to Avoid $\Delta Y = 1$ Neutral Currents, *Phys. Rev.* **D8**, 960 (1973).
158. *Particles, Sources and Fields*, Vol. II, Addison-Wesley, Reading, Mass., 1973.
159. Radiative Polarization of Electrons (with W.-Y. Tsai), *Phys. Rev.* **D9**, 1843 (1974).
160. A Report on Quantum Electrodynamics, in *The Physicist's Conception of Nature*, edited by J. Mehra (Reidel, Dordrecht, 1973), p. 413.
161. Spectral Forms for Three-Point Functions, *Phys. Rev.* **D9**, 2477 (1974).
162. Precession Tests of General Relativity — Source Theory Derivations, *Am. J. Phys.* **42**, 507 (1974).
163. Spin Precession — A Dynamical Discussion, *Am. J. Phys.* **42**, 510 (1974).
164. Photon Propagation Function: Spectral Analysis of Its Asymptotic Form, *Proc. Natl. Acad. Sci. U.S.A.* **71**, 3024 (1974).
165. Photon Propagation Function: A Comparison of Asymptotic Functions, *Proc. Natl. Acad. Sci. U.S.A.* **71**, 5047 (1974).
166. Interpretation of a Narrow Resonance in e^+e^- Annihilation, *Phys. Rev. Lett.* **34**, 37 (1975).
167. Source Theory Viewpoints in Deep Inelastic Scattering, *Proc. Natl. Acad. Sci. U.S.A.* **72**, 1 (1975).

168. Source Theory Discussion of Deep Inelastic Scattering with Polarized Particles, *Proc. Natl. Acad. Sci. U.S.A.* **72**, 1559 (1975).
169. Psi Particles and Dyons, *Science* **188**, 1300 (1975).
170. Resonance Interpretation of the Decay of $\psi'(3.7)$ into $\psi(3.1)$ (with K. A. Milton, W.-Y. Tsai and L. L. DeRaad, Jr.), *Phys. Rev.* **D12**, 2617 (1975).
171. Pion Spectrum in Decay of $\psi'(3.7)$ to $\psi(3.1)$ (with K. A. Milton, W.-Y. Tsai and L. L. DeRaad, Jr.), *Proc. Natl. Acad. Sci. U.S.A.* **72**, 4216 (1975).
172. Magnetic Charge and the Charge Quantization Condition, *Phys. Rev.* **D12**, 3105 (1975).
173. Source Theory Analysis of Electron-Positron Annihilation Experiments, *Proc. Natl. Acad. Sci. U.S.A.* **72**, 4725 (1975).
174. Casimir Effect in Source Theory, *Lett. Math. Phys.* **1**, 43 (1975).
175. Magnetic Charge, in *Gauge Theories and Modern Field Theory*, edited by R. Arnowitt and P. Nath (MIT Press, Cambridge, Mass., 1976), p. 337.
176. Classical and Quantum Theory of Synergic Synchrotron-Cerenkov Radiation (with W.-Y. Tsai and T. Erber), *Ann. Phys. (N.Y.)* **96**, 303 (1976).
177. Gravitons and Photons: The Methodological Unification of Source Theory, *Gen. Rel. and Grav.* **7**, 251 (1976).
178. Deep Inelastic Scattering of Leptons, *Proc. Natl. Acad. Sci. USA* **73**, 3351 (1976).
179. Deep Inelastic Scattering of Charged Leptons, *Proc. Natl. Acad. Sci. USA* **73**, 3816 (1976).
179a. Deep Inelastic Scattering of Polarized Electrons–A Dissident View, Talk presented at Symposium on High Energy Physics with Polarized Beams and Targets, Argonne Nat. Lab., August 22–27, 1976. (New York, 1976) pp. 288–305.
180. Nonrelativistic Dyon-Dyon Scattering (with K. A. Milton, W.-Y. Tsai, L. L. DeRaad, Jr., and D. C. Clark), *Ann. Phys. (N.Y.)* **101**, 451 (1976).
181. Adler's Sum Rule in Source Theory, *Phys. Rev.* **D15**, 910 (1977).
182. Deep Inelastic Neutrino Scattering and Pion-Nucleon Cross Sections, *Phys. Lett.* **B67**, 89 (1977).
183. Deep Inelastic Sum Rules in Source Theory, *Nucl. Phys.* **B123**, 223 (1977).
184. The Majorana Formula, *Trans. N. Y. Acad. Sci.* **38**, 170 (1977). (Rabi Festschrift).

185. Introduction and Selected Topics in Source Theory, in *Proceedings of Recent Developments in Particle and Field Theory*, Tubingen 1977 (Braunschweig, 1979), pp. 227–333.
186. New Approach to Quantum Correction in Synchrotron Radiation (with W.-Y. Tsai), *Ann. Phys. (N.Y.)* **110**, 63 (1978).
187. Casimir Effect in Dielectrics (with L. L. DeRaad, Jr. and K. A. Milton) *Ann. Phys. (N.Y.)* **115**, 1 (1978).
188. Casimir Self-Stress on a Perfectly Conducting Spherical Shell (with K. A. Milton and L. L. DeRaad, Jr.), *Ann. Phys. (N.Y.)* **115**, 388 (1978).
189. Introduction to Source Theory, with Applications to High Energy Physics, *Proceedings of the Seventh Particle Physics Conference*, (University of Hawaii Press, 1978), pp. 341–481.
190. Multispinor Basis of Fermi-Bose Transformation, *Ann. Phys. (N.Y.)* **119**, 192 (1979).
191. Relativistic Comets, *Kinam*, **1**, 87 (1979).
192. Thomas–Fermi Model: The Leading Correction, *Phys. Rev.* **A22**, 1827 (1980).
193. Thomas–Fermi Model: The Second Correction, *Phys. Rev.* **A24**, 2353 (1981).
194. New Thomas–Fermi Theory: A Test (with L. DeRaad, Jr.), *Phys. Rev.* **A25**, 2399 (1982).
195. Thomas–Fermi Revisited: The Outer Regions of the Atom (with B.-G. Englert), *Phys. Rev.* **A26**, 2322 (1982).
196. *The Statistical Atom: A Study.* University of Miami, P.A.M. Dirac Birthday Volume, 1982.
197. Quantum Electrodynamics, *J. Physique* **43**, 409 (1982).
198. Electromagnetic Mass Revisited, *Found. Physics* **13**, 373 (1983).
199. Renormalization Theory of Quantum Electrodynamics: An Individual View, in *The Birth of Particle Physics* (Cambridge University Press, 1983), p. 329.
200. Two Shakers of Physics, in *The Birth of Particle Physics* (Cambridge University Press, 1983), p. 354.
201. Statistical Atom: Handling the Strongly Bound Electrons (with B.-G. Englert), *Phys. Rev.* **A29**, 2331 (1984).
202. Statistical Atom: Some Quantum Improvements (with B.-G. Englert), *Phys. Rev.* **A29**, 2339 (1984).
203. New Statistical Atom: A Numerical Study (with B.-G. Englert), *Phys. Rev.* **A29**, 2353 (1984).
204. Semiclassical Atom (with B.-G. Englert), *Phys. Rev.* **A32**, 26 (1985).

205. Linear Degeneracy in the Semiclassical Atom (with B.-G. Englert), *Phys. Rev.* **A32**, 36 (1985).
206. Atomic-Binding-Energy Oscillations (with B.-G. Englert), *Phys. Rev.* **A32**, 47 (1985).
207. *Einstein's Legacy: The Unity of Space and Time*, Scientific American Library, Vol. 16 (1985).
208. Is Spin Coherence Like Humpty Dumpty? I. Simplified Treatment (with B.-G. Englert and M. O. Scully), *Found. Phys.* **18**, 1045 (1988).
208a. Hermann Weyl and Quantum Kinematics, in *Exact Sciences and Their Philosophical Foundations* (Verlag Peter Lang, Frankfurt, 1988).
209. Is Spin Coherence Like Humpty Dumpty? II. General Theory (with M. O. Scully and B.-G. Englert), *Z. Phys.* **D10**, 135 (1988).
210. Spin Coherence and Humpty Dumpty. III. The Effects of Observation (with M. O. Scully and B.-G. Englert), *Phys. Rev.* **A40**, 1775 (1989).
211. *Particles, Sources, and Fields* (3 volumes), Addison-Wesley, Redwood City, CA (1989).
212. A Path to Quantum Electrodynamics, *Physics Today*, February 1989. [Reprinted in *Most of the Good Stuff: Memories of Richard Feynman*, edited by L. M. Brown and J. S. Rigden (AIP, New York, 1993), p. 59.]
213. Cold Fusion: A Hypothesis, *Z. Nat. Forsch. A* **A45**, 756 (1990).
214. Nuclear Energy in an Atomic Lattice I, *Z. Phys.* **D15**, 221 (1990).
215. Anomalies in Quantum Field Theory, in *Superworld III, Proceedings of the 26th Course of the International School of Subnuclear Physics*, Erice, Italy, 7–15 August 1988 (Plenum, New York, 1990).
216. Phonon Representations, *Proc. Natl. Acad. Sci. USA* **87**, 6983 (1990).
217. Phonon Dynamics, *Proc. Natl. Acad. Sci. USA* **87**, 8370 (1990).
218. Reflecting Slow Atoms from a Micromaser Field (with B.-G. Englert), *Europhysics Lett.* **14**, 25 (1991).
219. Nuclear Energy in an Atomic Lattice, *Prog. Theor. Phys.* **85**, 711 (1991).
220. Phonon Green's Function, *Proc. Natl. Acad. Sci. USA* **88**, 6537 (1991).
221. Casimir Effect in Source Theory II, *Lett. Math. Phys.* **24**, 59 (1992).
222. Casimir Effect in Source Theory III, *Lett. Math. Phys.* **24**, 227 (1992).
223. Casimir Energy for Dielectrics, *Proc. Natl. Acad. Sci. USA* **89**, 4091 (1992).

224. Casimir Energy for Dielectrics: Spherical Geometry, *Proc. Natl. Acad. Sci. USA* **89**, 11118 (1992).
225. Casimir Light: A Glimpse, *Proc. Natl. Acad. Sci. USA* **90**, 958 (1993).
226. Casimir Light: The Source, *Proc. Natl. Acad. Sci. USA* **90**, 2105 (1993).
227. Casimir Light: Photon Pairs, *Proc. Natl. Acad. Sci. USA* **90**, 4505 (1993).
228. Casimir Light: Pieces of the Action, *Proc. Natl. Acad. Sci. USA* **90**, 7285 (1993).
229. The Greening of Quantum Field Theory: George and I, Lecture at Nottingham, July 14, 1993 (hep-ph/9310283).
230. Casimir Light: Field Pressure, *Proc. Natl. Acad. Sci. USA* **91**, 6473 (1994).

Compiled by K. A. Milton, 1978, 1994, 1995.

ABOUT THE EDITOR

Y. Jack Ng received his B.A. from Berkeley and his Ph.D. from Harvard. After postdoctoral studies at the Institute for Advanced Study, Princeton and the Stanford Linear Accelerator Center, he accepted a professorship at the University of North Carolina at Chapel Hill. A former Alfred Sloan fellow, Ng has been principal investigator of a US Department of Energy research grant since 1979. Reflecting the diverse interests of his mentor, Julian Schwinger, in physics, he has worked in a variety of fields, primarily particle physics and field theory, but also gravity, cosmology, and statistical physics.

www.ingramcontent.com/pod-product-compliance
Lightning Source LLC
Chambersburg PA
CBHW081203170426
43197CB00018B/2913